# Intelligent Systems Reference Library

Volume 263

The aim of this series is to publish a Reference Library, including novel advances and developments in all aspects of Intelligent Systems in an easily accessible and well structured form. The series includes reference works, handbooks, compendia, textbooks, well-structured monographs, dictionaries, and encyclopedias. It contains well integrated knowledge and current information in the field of Intelligent Systems. The series covers the theory, applications, and design methods of Intelligent Systems. Virtually all disciplines such as engineering, computer science, avionics, business, e-commerce, environment, healthcare, physics and life science are included. The list of topics spans all the areas of modern intelligent systems such as: Ambient intelligence, Computational intelligence, Social intelligence, Computational neuroscience, Artificial life, Virtual society, Cognitive systems, DNA and immunity-based systems, e-Learning and teaching, Human-centred computing and Machine ethics, Intelligent control, Intelligent data analysis, Knowledge-based paradigms, Knowledge management, Intelligent agents, Intelligent decision making, Intelligent network security, Interactive entertainment, Learning paradigms, Recommender systems, Robotics and Mechatronics including human-machine teaming, Self-organizing and adaptive systems, Soft computing including Neural systems, Fuzzy systems, Evolutionary computing and the Fusion of these paradigms, Perception and Vision, Web intelligence and Multimedia.

Indexed by SCOPUS, DBLP, zbMATH, SCImago.

All books published in the series are submitted for consideration in Web of Science.

Smaranda Belciug

# The Hospital Manager's Guide to Artificial Intelligence

## Concepts, Methods, and Techniques

Springer

Smaranda Belciug
Faculty of Science
Department of Computer Science
University of Craiova
Craiova, Dolj, Romania

ISSN 1868-4394 ISSN 1868-4408 (electronic)
Intelligent Systems Reference Library
ISBN 978-3-031-80313-0 ISBN 978-3-031-80314-7 (eBook)
https://doi.org/10.1007/978-3-031-80314-7



This Springer imprint is published by the registered company Springer Nature Switzerland AG
The registered company address is: Gewerbestrasse 11, 6330 Cham, Switzerland

*To Prof. Sally McClean,*

*Prof. Florin Gorunescu, and Prof. Emeritus Peter Millard*

*This book is as much yours as it is mine.*

# Foreword

This book is concerned with the use of artificial intelligence to provide advice and support for hospital managers. As in most aspects of modern life, huge amounts of data are now available to underpin management and decision making; however, to make sense of such vast resources, we need new technologies, particularly Artificial Intelligence (AI), underpinned by traditional statistical methods of data analysis. Such ideas are deep-rooted in Mathematics and help to ensure that the use of Artificial Intelligence is logical, principled, and trustworthy. This book provides a very readable, yet technical, discussion of the emerging area of intelligent healthcare in a comprehensive manner, ranging from high level hospital management and decision making to different hospital departments. These include important topics for the modern-day hospital manager who must find cost-effective ways of managing hospitals, while providing the best possible experience and outcomes for the patients.

Overall, while the book is necessarily technical in nature, is very readable with lots of interesting historical anecdotes, and can be enjoyed by students and experts from many areas of computer science, statistics and related disciplines, as well the primary audience of hospital managers and healthcare personnel, seeking to understand how these new technologies can benefit their work. Throughout there are many examples, web-links and illustrations that make it easier for the newcomer to understand the concepts, while the more experienced reader can enjoy the anecdotes, explanations and illustrations. Medicine and healthcare have made huge advances over recent decades, with many of these innovations being made possible or enhanced by Computer Science. However, such developments typically require additional resources and lead to increased costs. In such a climate, modern approaches, such as Artificial Intelligence, can offer assistance to the hospital manager by providing more targeted and optimised solutions.

The book begins with a gentle introduction to decision making for hospital managers, based on quantitative analysis. In management, such ideas are necessary to provide an efficient service while optimizing costs, revenue, and profits. Various statistical distributions are introduced and motivated by examples and plots. These ideas can be expanded to more sophisticated concepts such as statistical tests, and data gathering, particularly for a clinical department. In such situations, queueing models are of major significance, as the build-up of queues can lead to major inefficiencies, increased costs and distress to the patients. Another interesting area presented is the use of AI to avoid patient over-testing, over-treating, and over-resourcing. Then, in Pathology and Radiology departments, AI can leverage methods such as Neural Networks, particularly Deep Learning, for sophisticated images processing. A further topic discussed is the use of Internet of Surgical Things (IoST) to collect, combine and analyse data from different sources and devices, thus providing improved support, while Virtual and Augmented Reality can enhance this even further. The use of multiple AI models in disaster scenarios is also discussed.

This is a fascinating book, which succeeds in entertaining the reader as well as providing an excellent introduction to the use of Artificial Intelligence in hospitals. In addition, most importantly, it illustrates the discussion will a myriad of examples and situations where Artificial Intelligence can be brought to bear on hospital management to improve the quality and efficiency of hospital management and the patient experience. As such, it provides a very readable, but also technical, discussion, as well as motivating and inspiring the reader to learn more.

October 2024

Prof. Sally McClean
Ulster University
Belfast, UK

# Preface

In 2600 BCE the only facilities for the sick were temples dedicated to the "healing Gods", like Imhotep and Asclepius. The plans for the 5th century BCE "hospital" in Athens present a large room that held multiple beds for patients. Were the first hospitals built in Mesopotamia, or in the Buddhist monasteries in India and Sri Lanka in the 4th century BCE? The word hospital comes from Latin. As the Roman Empire fell, the Christian Church took the role of tending for the sick. Thus, hospitals started to be organized as wards. The belief was that if a sick patient was closer to the altar, he/she would heal faster. Also, the wards were shaped like a cross.

Let us fast forward over 4000 years. We arrive here, today. We live in an era where things change in a blink of an eye. We cannot keep up with the amount of data that is produced on a regular basis. Especially when it comes to healthcare. So, what is there to be done? Use Artificial Intelligence.

No matter who you are or why you are reading this book, no matter what your prior knowledge regarding mathematics and computer science is, this book is for everyone. It is written in a friendly manner. We will be your guide in this exciting journey of how to run a hospital using as counsellor Artificial Intelligence. We shall start by discussing the cookbook for decision making, followed by what can be improved in each department starting from clinical all the way to radiology and pathology. We will cover the inpatient and outpatient intelligent management; we shall see how the operation theatre becomes smart, and how we can improve the nursing station. Last, but not least, we shall cover the stock policy so that we reduce the costs, without losing quality in patient management.

By presenting concepts, models, and practical examples, we shall show you that Artificial Intelligence is nothing but mathematics with a high computational power.

I dedicate this book to the three mentors who have inspired me to write it and provided me guidance in this domain throughout the years: to my beloved father, Prof. Florin Gorunescu, and to my dear friend Prof. Emeritus Peter Millard, though you are no longer with me, your teachings inspired every word of this book. Your legacy lives on in the work of those you've mentored, including myself.

Last, but not least, to my mentor, Prof. Sally McClean, who inspires me to be better at my job, and to keep on hiking mountains! Thank you, Sally!

I want to express my respect and gratitude to Prof. Lakhmi Jain for offering me the opportunity to write this book and for his precious friendship. I would also like to thank Prof. Janus Kacprzyk for warmly sustaining this project.

Craiova, Romania
October 2024

Smaranda Belciug

# Contents

# Chapter 1
# Hospital Manager's Cookbook for Decision Making

**Abstract** The aim of this chapter is to provide a cookbook for the hospital manager for decision making. Before deciding, a manager needs to understand the theory behind problem solving and how to perform a quantitative analysis. To optimize costs, revenue, and profit, the manager needs to learn how to mathematically model the hospital's situation. After learning the theory, one needs practice, hence, practical examples are presented throughout the chapter.

## 1.1 Problem Solving

From the day we are born, till the day we die we learn and cope with managing different things. We manage our emotions, we manage our budget, we manage our time. If you stop from your daily rush, and think about it, everything is manageable. We manage life, and we manage death. Every decision that we take will have a direct impact on our lives. So how can we make a positive one on our life? We could just leave it all to chance, and hope for the best, but this is a long shot. We have little to no control on what is happening around us, so at least we should be able to control what we can. You cannot control what happens to you, but you can control how you react to it. When a problematic situation appears, two possibilities exist: either the adrenaline rush kicks in, and we find a solution, either it doesn't, and we find ourselves feeling lost. We cannot control the adrenaline rush, because it depends on how we are designed and built as a human being. So, what is there to be done? Hope for the best, but expect the worst. Prepare ourselves for potential problematic situations, and if and when they appear, apply the predetermined plan. This is why first responders spend hours and hours training for every possible situation that might appear.

*A little bit of history, anyone?*

Have you ever stopped and wondered how the healthcare system looked in Antiquity? If we go back in history, and wake up in Ancient Mesopotamia, we would see the temples of Gula, the Sumerian goddess of medicine, that served as hospitals, [1].



S. Belciug, *The Hospital Manager's Guide to Artificial Intelligence*, Intelligent Systems Reference Library 263, https://doi.org/10.1007/978-3-031-80314-7_1

Within their high walls, patients were healed by dogs licking their wounds. After the wound was cleaned, different herbal ointments were applied on it. The medical pioneers started in Mesopotamia. As any other doctor, Gula did not work alone in the process of healing. She was helped by her spouse Pabilsag, and her children Damu, Nianzu, and Gunurra. All humankind is familiar with the insignia of the medical profession, the rod tangled with a serpent, but not everybody knows that the sign is originated with Ninazu who was associated with serpents, the underworld, the healing process, and transformation, [2, 3]. The doctors of Ancient Mesopotamia were the agents that were used by above mentioned deities to manage the healthcare system. Just like today, the first step in healing someone, was to diagnose them. The cause was always a sin that has been committed by the patient. Since the patient was touched by the hand of a certain God, the cure was in the will of the gods. To be cured, a patient had to confess and commit to be a better person in the future. If the patient died, then chances were that he/she offended another god.

In a Mesopotamian hospital there were two kinds of doctors: the *asu*, which was a physician that performed medicine empirically, and the *asipu*, which was a healer that used magic to treat patients. Let us see for instance how birth was managed during that time: while the *asipu* prayed to the gods and made incantations to scare off the demons (the demon Lamashtu was to blame for newborn death), the *asu* eased the labor pains with different herbs. The actual birth was handled by a midwife.

The Mesopotamian healthcare system was organized as follows: physicians performed surgeries and treated patients within the temples, but they also made house calls. It is thought that physicians were trained in the city of Isin, after which they were distributed to different temples in various cities. Kings and important persons had their own doctors. No evidence of private practices was found. To recognize a doctor, you just had to take a good look at his head: if she/he was shaved (sometimes only on the left side), that certified that she/he belonged to this profession. Have you noticed that we have used the "she" pronoun? That is because there were a lot of female physicians in the Early Dynastic Period of Sumer (2900–2334, B.C.).

What about the healthcare resources in Ancient Mesopotamia? Well, the doctors carried their instruments and a portable bed with them. Even back then, sickness was associated with filth, and since the poor would sleep on mats directly on the dirty floor, it was believed that if you elevated the sick person onto a bed, it would heal her/him quicker. Since, the gods made people sick and also healed them, doctors were not held accountable for their results. There was only one exception: surgery. If the doctor performed surgery and had a poor outcome, one or both her/his hands would have been amputated. This was the 218 mandates of the Hammurabi Code. Unfortunately, since dissection was forbidden, little to no knowledge regarding the physiology or anatomy was known, [4].

The modern healthcare procedures were determined by the Ancient Egyptians. Just like in Ancient Mesopotamia, in Ancient Egypt, doctors work side-by-side with priests and healers. In this manner, by combining medicine with magic and religious rituals, the healing process became holistic. Diagnosing patients was a little bit far from personalized medicine, but not as far as you might think. Besides examining the patient's body, measuring her/his pulse, and taking account of the symptoms, doctors were paying attention to the patients' dreams and omens. Using prayers as well as remedies from herbs, Egyptian doctors were practicing medicine. All the remedies were written down in medical papyruses, such as Ebers Papyrus or Edwin Smith Papyrus. The Egyptian healthcare system was based on two fundamental things: cleanliness and hygiene. Physicians knew that in order for a wound not to become infected, one must keep it clean. They paid special attention to the patient's diet and nutrition, since their belief was that certain food helps patients recover faster. The phrase "mind over body" meant something in Ancient Egypt also. Physicians believed that positive thinking promoted healing by uplifting the patient's spirit.

Imhotep (2667–2600 B.C.) was the first physician. He designed the Step Pyramid of Djoser at Saqqara. He was the first to argue that diseases were not the punishments from the gods. In Ancient Egypt, women were allowed to be doctors. Merit-Ptah was the first female doctor known by name, but some evidence points to the fact that the medical school at the Temple of Neith in Sais, was run by a woman in 3000 B.C.

The Egyptian healthcare system was organized as follows: the doctors were *wabau* and could have her/his specialty, and were expected to bathe frequently, or could be *swnw*, that is general practitioners, or *sau*, the physicians that were specialized in magic. Besides them, they had midwives, masseurs, attendants, nurses, and clairvoyants. Just like in Ancient Mesopotamia, doctors did not attend births. Midwives were female relatives, neighbors, or friends, and were not seen as medical professionals, since they had no medical training. Nurses on the other hand were both male and female, and they had professional training. There were two types of nurses: wet and dry. The wet nurse was more important than the dry nurse, because they assisted women in childbirth, and in the case the mother died, the wet nurse would continue to take care of the infant. The dry nurse assisted in different procedures and was thought of as being interconnected to the divine.

The Egyptian god of magic and medicine was Heka. Heka is visualized carrying a staff tangled with two serpents. From Egypt, the symbol passed onto the Greeks, that associated the tangled serpent with Asclepius, their god of healing. Nevertheless, keep in mind that the medical profession image comes from Ninazu, son of Gula. Besides Heka, there were other deities that delt with healing: Sekhmet, Sobek, and Nefertum. Each one of them was invoked in certain medical procedures. For instance, Sobek, the crocodile god was invoked in surgeries and invasive procedures. Egyptian medical practice was codified by Hippocrates (460–370 B.C.).

Texts written by Homer and Hesiod, depict Asclepius, the Greek god of healing. Son of Apollo, but resembling remarkably with Zeus, looking all mature, fatherly and bearded, Asclepius seemed more human than deity. He was seen more approachable to humans. He carried a walking stick, and kept near him a dog, and that's right, you have guessed correctly, a serpent.

As Asclepius' cult spread throughout Greece, more and more sanctuaries started being dedicated to healing. The formal hospitals started to be established. The first sanctuary was established in Trikka. During the fifth and fourth centuries B.C. more and more Asclepius sanctuaries appeared in Athens, Corinth, Sicyon, Argos, Sparta, Messene, Megalopolis, Parosm Crete, Pergamum (Asia Minor), Aegina, Alexandria (Egypt), Cyrene (Lybia), and even in Rome. It was the time, when every big town had a healthcare facility for its residents and nearby neighbors. Asclepieia included hospitals, nursing-homes, religious worship places, as well as medical schools. Hippocrates' followers taught medicine at Kos, and Galen was trained at Pergamum, before becoming Marcus Aurelius private physician.

Multiple gods were worshiped as Asclepieia, including Apollo, Artemis, and Asclepius' children Machaon, Podalirius, and Hygieia. In such a healthcare facility, we would find altars and temples, and the Abaton. The Abaton was a building in which the patients would undergo some sort of incubation, spending the night praying so that the god would come to them in a dream and suggest a course of treatment. Besides these, we would find baths, hostels, dining rooms, stadium, palaestra, a large gymnasium, and a theatre. Near the Abaton, we would see the Tholos, a circular structure in which snakes would slither freely. In general, the healthcare facilities were built near springs. The patients were purified first, before the incubation process, [5].

In India, starting with the sixth century B.C., the Buddhist religion made possible the creation of the monastic system, that help develop healthcare facilities in and near these monasteries. Evidence of the first nurse have been found here. In 431 B.C. the first Sri Lankan hospitals appeared. It is impressive to know, that the great Indian king Asoka has built hospitals for people as well as for animals during the third century B.C. As Buddhism spread, hospitals started to appear in China and Southeast Asia. An interesting fact is that in the Jewish history there is no mention of a healthcare system. In the Bible, only private homes are referred as healing places, see the Good Samaritan parable, Lazarus, the centurion's child, etc., [6, 7].

The hospitals that we are familiar with, have been developed at the end of the first millennium B.C. Because we are aiming to draw a parallel between the way a hospital is organized today and how it has been organized throughout history, we can think of the *iatreia* (Greek), or the *taberna medica* (Roman) as short-term care. Iatreia and taberna medica were the doctor's consultation room. The long-term care took place in private homes, [8].

As the Roman Empire started to expand, it was impossible to send back home the sick and wounded soldier. Hence, military hospitals were built. There were two types of military hospitals: *valetudinaria* fort hospitals that were permanent, and field hospitals which were temporary during the active campaigns, [9–11]. Evidence of the first valetudinarium were found near Vienna, at Carnuntum, and date from the first century AD. The larger valetudinarium had four wings and a central courtyard. The number of wards depended on the number of centurions in the legion. Each ward had maximum 3 beds and were very private. Besides this, they had treatment rooms, staff rooms, a dispensary, latrine, washing and cooking accommodations. The valetudinarium had medical staff and supply lines, that were marked as high priority. There is no evidence that the valetudianrium played a role in establishing public hospitals. For the general population the healing process was done in households.

In Christian times, hospitals provided shelter and healthcare for sick pilgrims, e.g. St Bartholomew Hospital, [12]. Infirmaries were healthcare facilities that were built and run inside monasteries. There were designed for providing care for the monks. As time passed, in the infirmaries there were allowed civilian patients. Different from Roman times, when economic factors played a huge role in building healthcare facilities, these hospitals were supported by Christian charity, [10].

One of the most impressive healthcare facilities was built by St. Basil of Caesarea (Cappadocia). His hospital had the number of wards equal to the number of known diseases. They had a ward for lepers, hospices, for travelers, for the poor, and an industrial school. After the plague of 375, at Edessa, St. Ephraim build a 300 bedded hospital. Lady Fabiola built the first hospital in Rome, in 390. Following her example, ladies Pulcheria and Pauline built hospitals in Constantinople and Jerusalem. The Fall of Rome in 476 brought the world in the Dark Ages, when Europe's development stopped.

Fast-forwarding over 300 hundred years, we arrive in the time of Charlemagne (747–814) that promulgated a decree that attached a school, a hospital, and monastery to every cathedral, [6, 7]. During the Middle Ages several monastic hospitals were built. Here we mention: Merida, Spain (580), St. John's Hospital, Ephesus (610), Pantokrator, Constatinople (seventh century), Hotel Dieu, Paris (651), St Maria della Scala, Siena (898), [6]. After the tenth century, monks were banned to practice medicine outside the monasteries, increasing the role of physicians. Hospitals started to be specialized. There were leper houses, maternity hospitals, and homes and hospitals for the elderly, [13].

The Islamic revolution started in 632 and ended in the thirteenth century. This was the Golden Age of Islam. The Greek medical dogma was studied and translated in Arabic. Important contributions to the medical domain were added by great physicians such as Rhazes (866–832), Albucasis (936–1013), Avicenna (980–1037), Avenzoar (1091–1162), Averroes (1126–1198), and Maimoides (1135–1204), [14]. Impressive hospitals, *bimaristans*, were built at Baghdad, Cordoba, Damascus, Sevilla, Cairo, and Bokhara. There have been reported 60 hospitals in Bagdad, and 50 in Cordoba. The most outstanding hospital was the Mansuri hospital in Cairo. It had four courts with water fountains, separate male/female wards, wards for different diseases, lecture halls, dispensary, and out-patient department, a library, and a chapel. Special fever wards were designed to be cooled by fountains. The patients were entertained by artists such as musicians and storytellers. At discharge, each person received money so that they could manage their living until they could work again, [14–16].

For 5000 years, humankind tried to prepare itself to deal with the sick and injured, whether it meant praying to the right god, preparing different ointments, or arranging healthcare facilities, [17]. You can easily see that problem solving has been part of our lives since forever.

But what is problem solving? The easiest definition would be: the strategy and roadmap that one has to build in order to change the present state of affairs into a desired one. We use problem solving to understand our environment, to identify what we need or want to change, and how to do that. Problem solving is the base of evolution. It is what makes us improve, and become a better version of ourselves, while building a better version of our world. We start problem solving since the day we are born. We learn to eat, talk, walk, etc. As we become older the problems that we need to solve become harder.

There is a saying: if it isn't broke, don't fix it! This is one way to look at a situation. The other is: there is always room for improvement. Obviously, problem solving comes with its share of risks. We need to learn how to address risks, to develop awareness skills that allow us to foresee the moment when something might break (based on past experience and current situation). If we can anticipate things that might happen, we can take action now and influence the likelihood of that event to occur, and in the case, it does indeed occur, we might alter its impact.

We live in a world where everything is interconnected and dynamic. Even if we oppose to change, somebody or something will change and thus change our environment. The interdependencies create a force that makes us continuously adapt and improve. Problem solving is not only about fixing or preventing bad things from happening. It is about seizing opportunities, creating innovation, and improving our lifestyle.

The problem-solving process involves the following steps presented in Figs. 1.1 and 1.2.

Steps 1 through 5 represent the *decision-making* process. The first step of a decision process is to identify what the problem is. Once you find the problem and understand what you must do to fix it, you need to define it in mathematical terms. When facing a problem, you might find yourself facing multiple potential solutions. Some of these solutions are better than others. It is up to you to determine criteria that help you evaluate them and determine which is the best one for a certain situation.

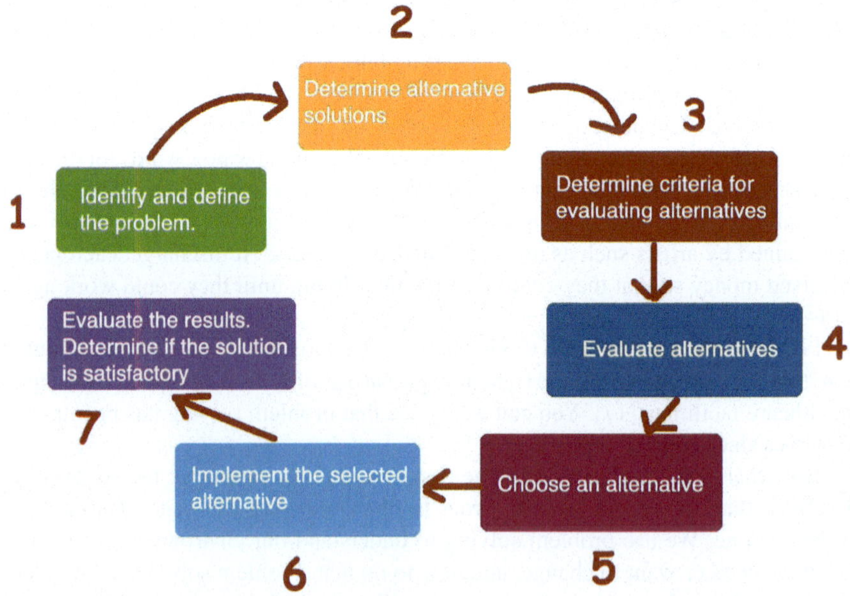


**Fig. 1.1** Steps 1–5 of the problem-solving process

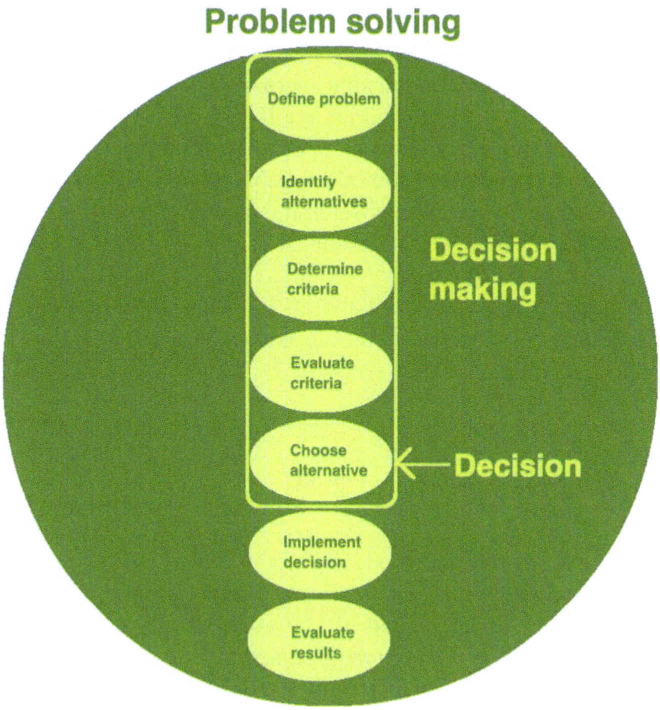


**Fig. 1.2** The problem-solving process

After you decide which alternative suits the problem, you choose it, thus you are making a decision.

Let us consider an example. You are the manager of private practice. There is an opening in your clinic, and you need to occupy that position. Hence you organize a job competition, and two candidates apply for it. Your problem is that you want to fill that position with a doctor. Once you have defined this problem, you look at the potential solutions: the doctor's CV's and demands. The first doctor is an unexperienced one that asks for 35 € per hour and can consult in that time 3 patients. The second is an experienced doctor that asks for 50 € per hour and can consult in that time 4 patients. Therefore, we are presented with the following alternatives:

1. Give the job to the unexperienced doctor.
2. Give the job to the experienced doctor.

Now, we have arrived at step 3, and we need to determine which criteria or criterion to use to make the best decision. If we are aiming at spending less money, then the smallest price per hour should be the criterion, hence the choice would be hiring the unexperienced doctor. This is the case of *single-criterion* decision problem. In general, we do not encounter this type of problem in real case scenarios. Hence, let us add two other criteria: the number of patients seen per hour and the doctor's experience. Now, we are dealing with a *multicriteria* decision problem.

Step 4 involves evaluating the potential alternatives to our problem. Here we must take into account the priority of each criterion. If their weight is equal, then it is easy to see that the second doctor fulfils two out of three, so we should consider him/her as our first choice. If the criteria do not weight equally, then the problem changes. We shall deal with this situation later on in this book. Once we have evaluated the criteria, and we are satisfied with the result, we are ready to make a *decision*. With this, the decision process is over.

To get to better understanding of the art of management science in decision making, we will try to provide different examples, that a hospital manager might encounter in everyday life.

## 1.2 Quantitative Analysis

Quantitative comes from the Latin *quantitas* or *quantitativus* in Medieval Latin. In the early sixteenth century, it meant *"having magnitude or spatial extent"*. Analysis comes from the Greek *analuein*, and its Oxford dictionary definition is *"detailed examination of the elements or structure of something"*. So, in other words when we are referring to the quantitative analysis of a problem, we actually mean breaking down that problem into several components.

The decision process can be quantitative or qualitative. The latter is based on the hospital manager's intuition, judgement, and experience. It is more instinct, rather than science. This works if the manager has experience, and the problem is simple enough. If the manager has encountered in the past problems of the same nature and structure as the present one, then he/she can use his/her experience and solve it using qualitative analysis. However, if the manager hasn't got enough experience, or the problem is more complex, and he/she has never dealt with something similar before, then quantitative analysis is the answer. Truth to be told, we believe that no matter the level of experience, one should always use quantitative analysis, and not just rely on experience or intuition.

When we are using quantitative analysis, we are looking at the data, and we are concentrating our attention on facts. We use the data to build mathematical models that describe our objectives, the constraints we are facing, or any other relationship between the data. Using different quantitative methods, we make a recommendation.

We use quantitative analysis to evaluate, to measure, to predict. In order to do so, we need to transform raw data into quantitative data. This is done through data engineering (DE). DE involves collecting, organizing, and processing information. Management science is similar to cooking. You should think of the data as ingredients. These ingredients are found in different places: the eggs and butter are in the fridge, the vegetables and rice in the pantry, and the oil and salt in the cupboard. This is data organization or data structuring. Obviously, before cooking a meal we need to prepare the ingredients, or to process them: we wash the vegetables, sift the flour, or boil the eggs. In DE the processing step is called Extract, Transform, and Load or ETL. The recipe represents the documentation on how to use the data, and the baking is Exploratory Data Analysis or EDA. At this point, the data is ready to be served as input for an Artificial Intelligence (AI) system.

Just like analysis, data comes in two forms: quantitative and qualitative. Before starting any process, we should clearly define what type of data we are dealing with, because if we apply the wrong statistical tool the obtained results would be nonsense. All measurements in science can be done using four scales: nominal, ordinal, interval and ratio variables, [18, 19]. This scale was invented by Stanley Smith Stevens, the founder of Harvard's Psycho-Acoustic Laboratory. Initially, the scale was developed somehow empirically, the mathematical rigor appearing later in the studies of Theodore Alpen, Louis Nares, and R. Duncan Luce. Not everybody accepted this scale, some scientists opposed it, [20].

A qualitative feature is either *nominal* or *categorical*, and the most important aspect is the fact that there is no natural order between the categories it might belong to. For example, the ward a patient is admitted to is such a variable: pediatrics, surgery, cardiology, etc. The type of surgery is another example of qualitative variable (appendectomy, urological surgery, coronary artery bypass, cesarian section, etc.). You can see that there is no order whatsoever between the categories. For computational purposes, the values can be coded using numbers (0, 1, 2, ... etc.), but these values cannot be used to compute descriptive statistics parameters such as the mean, standard deviation, etc., because the obtained results would complete nonsense. A variable that can be ordered is an ordinal variable. For example, the short stay department, medium stay department, and long stay department.

The third type of variable that we are going to discuss is the *interval*. The values of an interval are ordered, and the differences between two values make sense (i.e. the difficulty grading of a procedure). The *ratio* variable is ordered also. If a ratio variable has the value 0.0, then it does not exist. An example of a ratio variable is the death-to-case ratio, i.e. the number of deaths attributed to a particular disease during a period of time.

Sometimes is hard to know what you can, or you cannot compute, if you are dealing with a certain type of variable. Therefore, we have provided a cheat sheet in Table 1.1, that you can use whenever you need.

Quantitative variables are *numerical*, and they are divided into *discrete* and *continuous*. Since they are numerical, one can compute both the mean, median, or standard

**Table 1.1** Nominal, ordinal, interval, and ratio variable statistic parameters

| Type | Nominal | Ordinal | Interval | Ratio |
|---|---|---|---|---|
| Frequency distribution | Yes | Yes | Yes | Yes |
| Median/percentiles | No | Yes | Yes | Yes |
| Add/substract | No | No | Yes | Yes |
| Mean, standard deviation | No | No | Yes | Yes |
| Ratio/coefficient of variation | No | No | No | Yes |

deviation. A counting process is described by *discrete* data (i.e. the number of days that a patient has spent, in the ICU, the number of surgeries that a doctor has performed during a period of time, etc.). *Continuous* data is obtained by measurement (i.e. the cost of a procedure, the profit, etc.). If things weren't complicated enough, sometimes, a discrete value can be interpreted as continuous (i.e. length of stay). At the first glance, this doesn't seem such a big problem, but in fact it is. Why? Because depending on the data type, you must choose the correct distribution for the model. In general, if the data is discrete, then you can choose between the *Binomial* distribution and the *Poisson* distribution; if it is continuous then you choose between the *Gaussian*, *Lognormal* distribution, or $\chi^2$ distribution.

Before we dive into discussing each distribution, we need to explain some statistical concepts such as:

- the *Probability Density Function* (PDF) computes the likelihood of a certain observation in a given distribution (Fig. 1.3):
- the *Cumulative Density Function* (CDF) computes the cumulative likelihood of the current observation and all prior observations. Its domain ranges between 0 and 1 (Fig. 1.4).

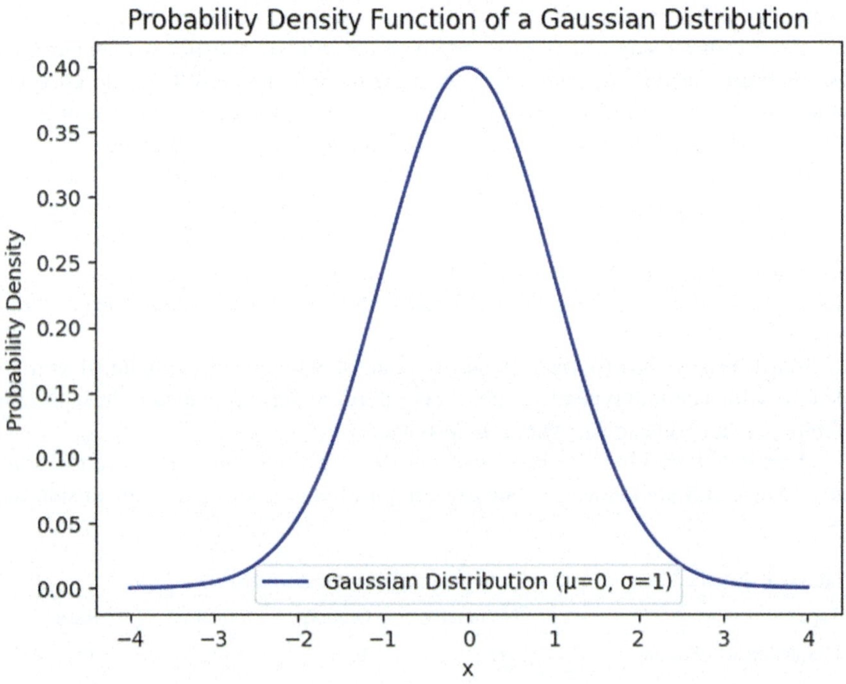


**Fig. 1.3** Gaussian PDF

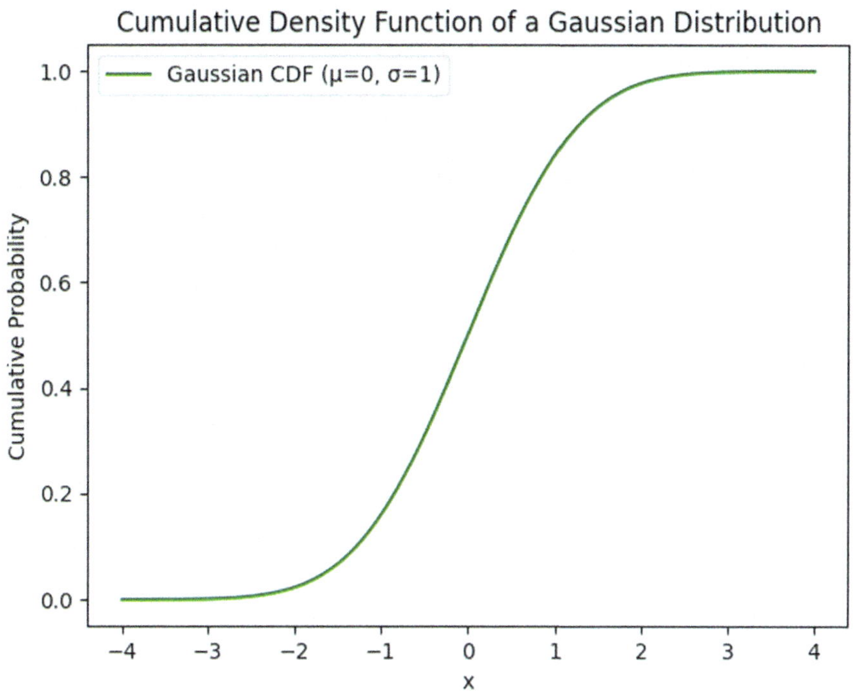


**Fig. 1.4** Gaussian CDF

- The *Probability Mass Function* (PMF) computes the probability of a discrete random variable to equal a given value. There should be noted that we can use the PMF only for discrete distributions (Bernoulli, Binomial, Geometric) (Fig. 1.5).

***Binomial distribution***

The Binomial distribution represents the probability of an experiment to be a success or a failure (e.g. the procedure has succeeded, and the patient lived, or it was a failure, and the patient died). The formula is:

$$b(x; n, p) = \binom{n}{x} p^x (1-p)^{n-x}, x = 0,1, \ldots, n$$

having the binomial coefficient:

$$\binom{n}{x} = \frac{n!}{x!(n-x)!}.$$

We translate the above formula into words as: the probability of the occurrence of $x$ successes is $p^x$, whereas the probability of occurrence for $n - x$ failures is

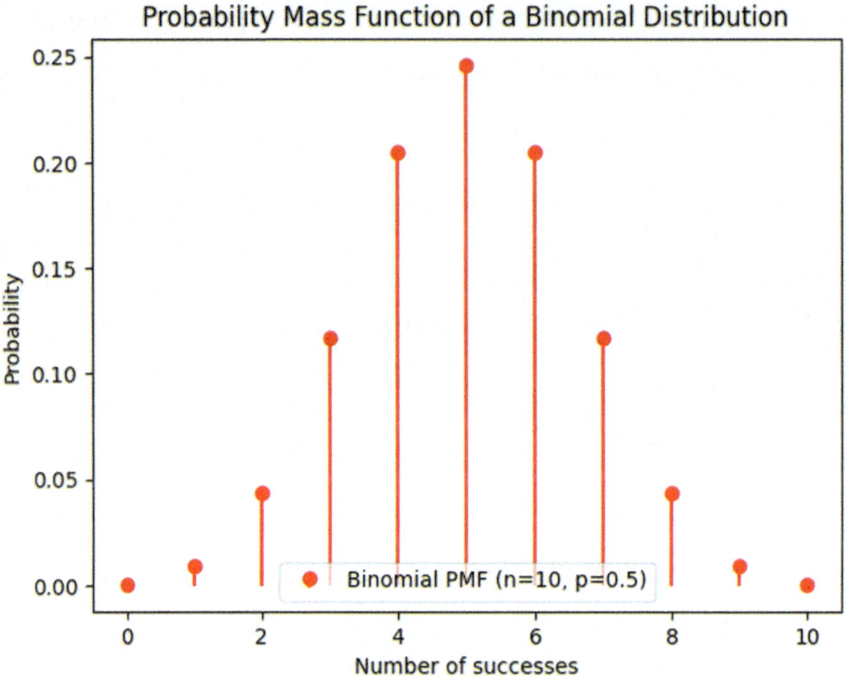


**Fig. 1.5** PMF for binomial distribution with $p = 0.5$

$(1-p)^{n-x}$. $n$ is the number of trials, and $x$ is the number of successes that appear randomly. $\binom{n}{x}$ represents the number of possibilities of how the successes will be distributed, [21–23]. The binomial distribution histogram and function plots are depicted in Figs. 1.6 and 1.7.

You might notice that the binomial distribution resembles a little bit to the geometric distribution, but pay attention, the geometric distribution describes the number of trials we need to wait before a success is observed. Before using the binomial distribution, we need to check the next four prerequisites:

1. All $n$ trials are independent.
2. The number of trials is fixed.
3. We can classify the trial's outcome either as a success or a failure. There is no third option.
4. The probability $p$ of a success is the same for each trial.

In a binomial distribution we have the following formulas for computing the mean ($\mu$), variance ($\sigma^2$), and standard deviation ($\sigma$):

$$\mu = np$$

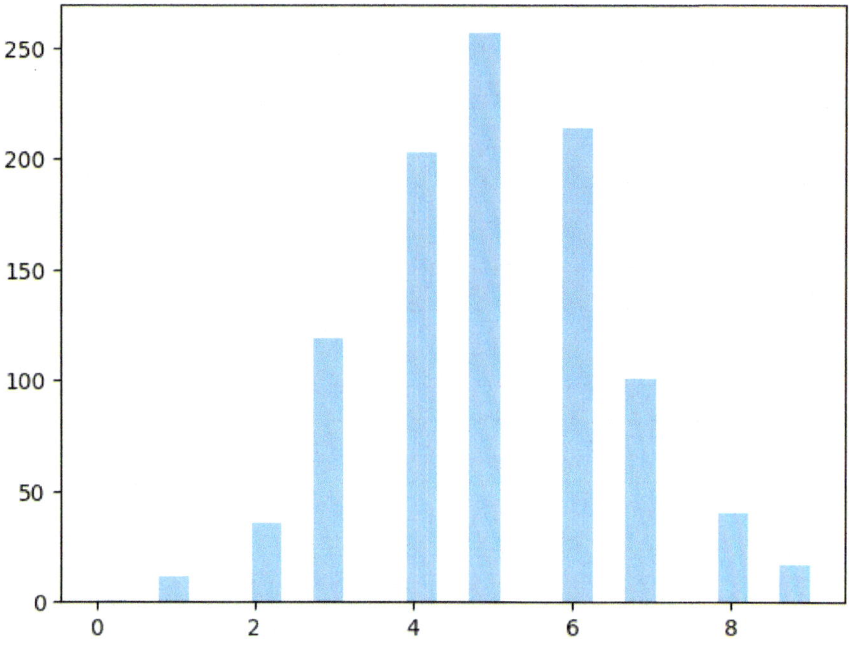


**Fig. 1.6** Binomial distribution histogram

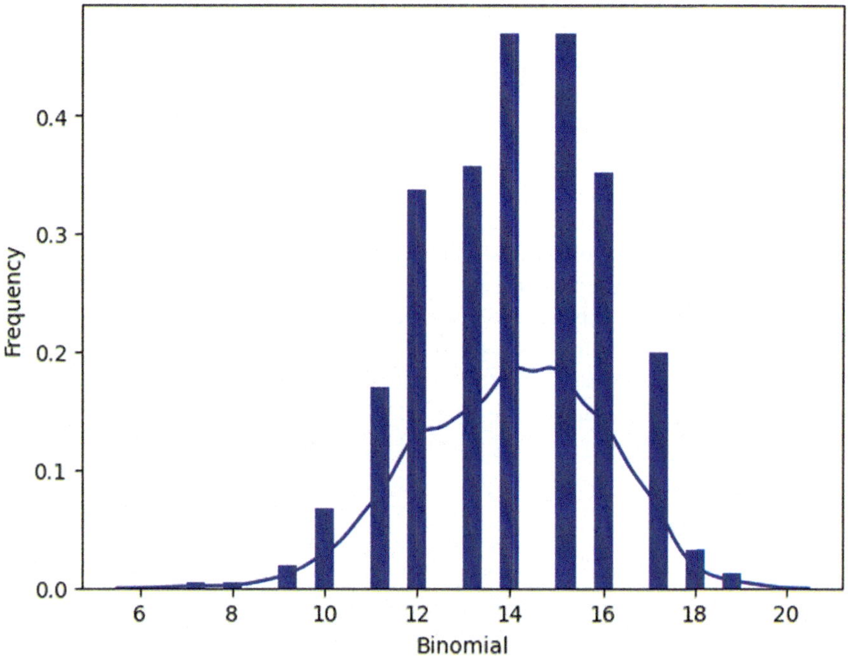


**Fig. 1.7** Binomial distribution function plot

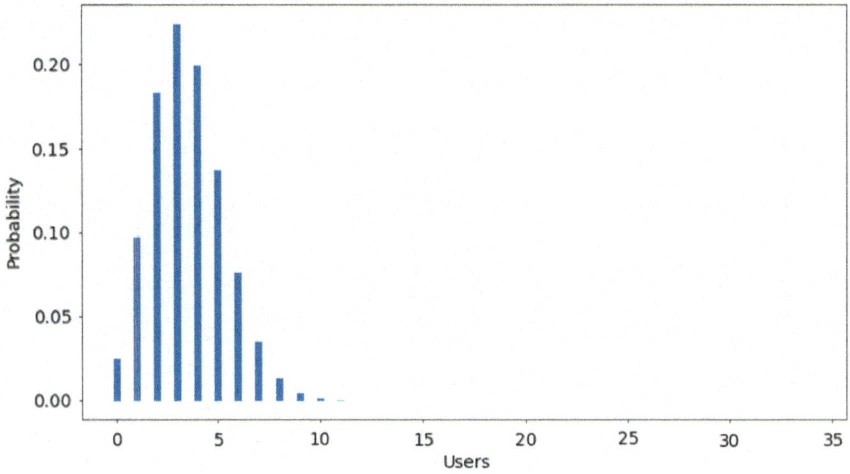


**Fig. 1.8** PMF $p = 0.1, n = 35$

$$\sigma^2 = np(1 - p)$$

$$\sigma = \sqrt{np(1 - p)}$$

Let us consider the following example: what is the probability of 3 patients out of 35 that came to the emergency room (ER) to be admitted that day? First, we shall plot the graph of having 35 patients visiting the ER, and the probability of them being admitted to the hospital that day (Fig. 1.8).

To answer the above question, we can look at the above plot, and see that it is around 20%, but if we want a more precise value, we can compute it, and the value is 0.224.

Let's see what is the probability of hiring 2 nurses out of 47 candidates, if previous information tells us that, in general, only 1 nurse is hired out of 47, taking into account their skills and experience versus the requested skills and experience? Using the binomial distribution the answer is 0.172.

One thing that needs to be noted is the fact that if we have a large enough sample $n$, that makes both $np$'s and $n(1 - p)$'s values at least 10, then the probability of success $p$ is approximately Normal. Figure 1.9 presents the Binomial versus Normal distribution plot.

***Poisson distribution***

We use the Poisson distribution, if we are planning to estimate the number of events in a large sample data over a period of time. For example, the number of people that suffer a stroke and are admitted to the Intensive Care Unit (ICU) in a month. The strokes occur over time with a fixed frequency, but each stroke occurs independently

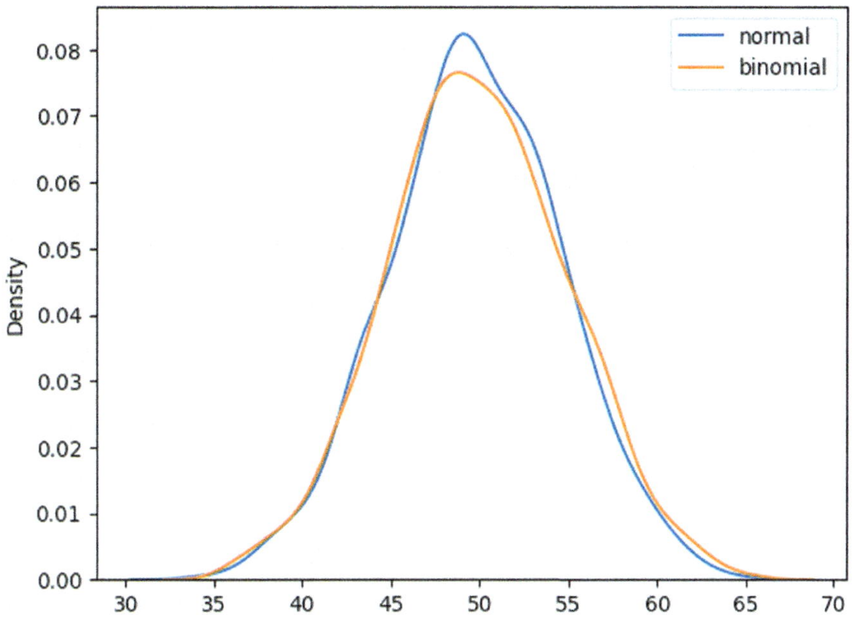


**Fig. 1.9** Binomial versus Normal distribution for large enough samples

from the rest, and totally random. For us to be able to apply Poisson, we need the events to be independent, their average rate per time to be constant, and not to have two events occurring at the same time. Using the Poisson distribution, we can predict the probability of an event or multiple events to happen in a predetermined period of time, taking into account how often that event usually happens. The average rate of occurrence per time is denoted by $\lambda$. Using $\lambda$, we can define the probability of observing an exact number of occurrences of a certain event over a period of time.

The following variables follow a Poisson distribution: the number of patients that come to the ER, or the number of patients that call to make a doctor's appointment. For example, if we want to know how many people will call to make an appointment at a private practice in the next 60 min, we can use the Poisson distribution and the Probability Mass Function to model it as follows:

$$P(X = k) = \frac{\lambda^k e^{-\lambda}}{k!}.$$

- The probability of having 52 people calling to make an appointment in the next 60 min, knowing that the average number of calls is 40, is 0.0106.
- The probability of having over 52 people calling to make an appointment in the next 60 min, knowing that the average is 40, is 0.028.

Let us see what is happening if for the same example of hiring 3 nurses out of 47 knowing that $p = 0.02$, we use the Poisson distribution and the Binomial distribution? If we use the Poisson distribution, we will get 0.054, and for the Binomial distribution we will get 0.053.

You can see that it is not an easy task to decide which distribution to choose between Poisson and the Binomial. In Fig. 1.10 we present the Poisson, Binomial and Normal distribution graphs. We can see that they resemble each other a lot. So, what is there to be done? We need to really pay attention to the data we are using as input, and also to the output data. For instance, if we have knowledge of an average probability of an event to happen in a given period of time, and we need to compute the probability of an event to happen in a period of time, or to compute the number of events that might occur in that period, then we need to choose Poisson distribution. Otherwise, if the input is an exact, not an average, probability and we need to compute the probability of an event to occur in an exact number of times out of $x$, (i.e. 70 out of 90), then we need to choose the Binomial distribution, [24–26].

The Poisson Distribution Probability Mass Function is presented in Fig. 1.11.

***Gaussian or Normal Distribution***

The most common distribution used in practice is the Gaussian or Normal distribution. In literature it is also called the Bell distribution due to its bell-shaped symmetric, unimodal curve, [27–33]. The distribution is noted $N(\mu, \sigma)$ because it depends on two parameters: the mean, $\mu$, and the standard deviation, $\sigma$. If $\mu = 0$ and $\sigma = 1$, then we are dealing with the *Standard Normal* distribution. The density and distribution functions are:

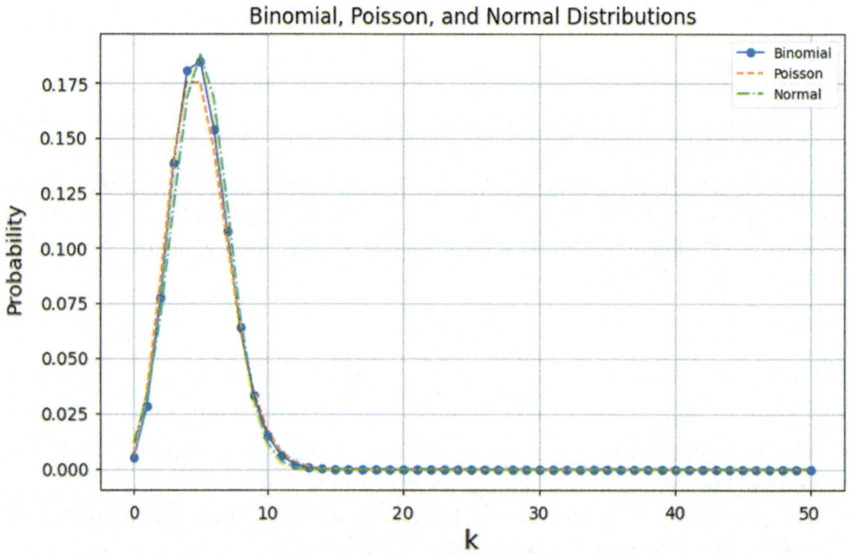


**Fig. 1.10** Binomial, Poisson, and Normal distribution graphs

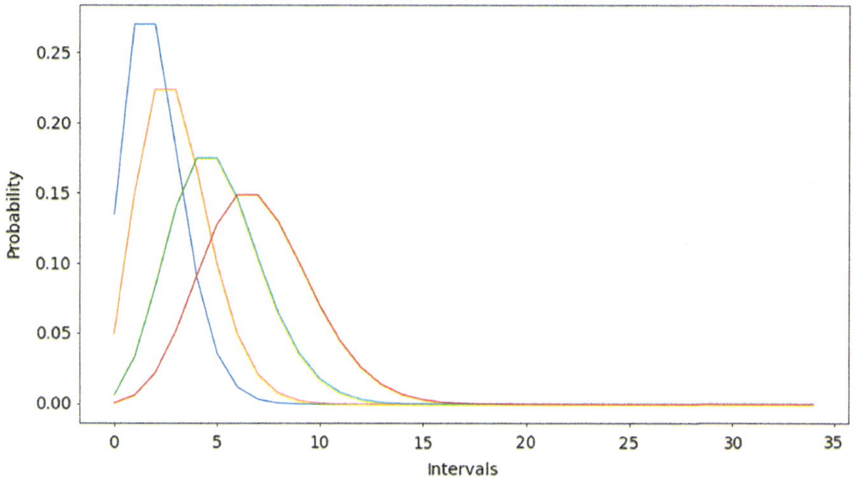


**Fig. 1.11** Poisson Distribution Probability Mass Function plot for different values

$$f_X(x) = \frac{1}{\sigma\sqrt{2\pi}} \cdot e^{-\frac{(x-\mu)^2}{2\sigma^2}}, -\infty < x < \infty,$$

$$F_X(x) = \frac{1}{\sigma\sqrt{2\pi}} \cdot \int_{-\infty}^{x} e^{-\frac{(t-\mu)^2}{2\sigma^2}} dt, -\infty < x < \infty.$$

The Gaussian distribution graph is plotted in Figs. 1.12 and 1.13. In Fig. 1.12 we have plotted multiple Gaussian distributions that have the mean$\mu = 0$, and different values for the standard deviation$\sigma \in \{1, 1.5, 2, 2.5, 3\}$. We can see that the graph is symmetric to the line$x = \mu$, line which admits a maximum point. In Fig. 1.13, we present the quartiles, and the inflection points at $\mu - x$ and $\mu + x$ for the Standard Normal distribution.

By looking at the Gaussian Bell, we can draw some conclusions:

- 95% of the data points lie between the lines $x = \mu - 2\sigma$ and $x = \mu + 2\sigma$, which represents the 95% confidence interval.
- 68% of the data points lie between the lines $= \mu - \sigma$ and $x = \mu + \sigma$.
- 99.7% of the data point lie between the lines $= \mu - 3\sigma$ and $x = \mu + 3\sigma$.

In Figs. 1.14 and 1.15 we depict how a Standard Normal distribution looks like versus a non-Standard Normal distribution plotted in 3D together with the PDF. The differences can be spotted on the *Ox* and *Oy* axis: in the standard case there are circles, and on the non-standard there are ellipses.

Let us take the following example: suppose we have admitted to the ICU a child that suffered a severe asthma crisis. The healthcare costs reported for this child were 6200 € for 4 days. We are interested in finding out how many children that suffer

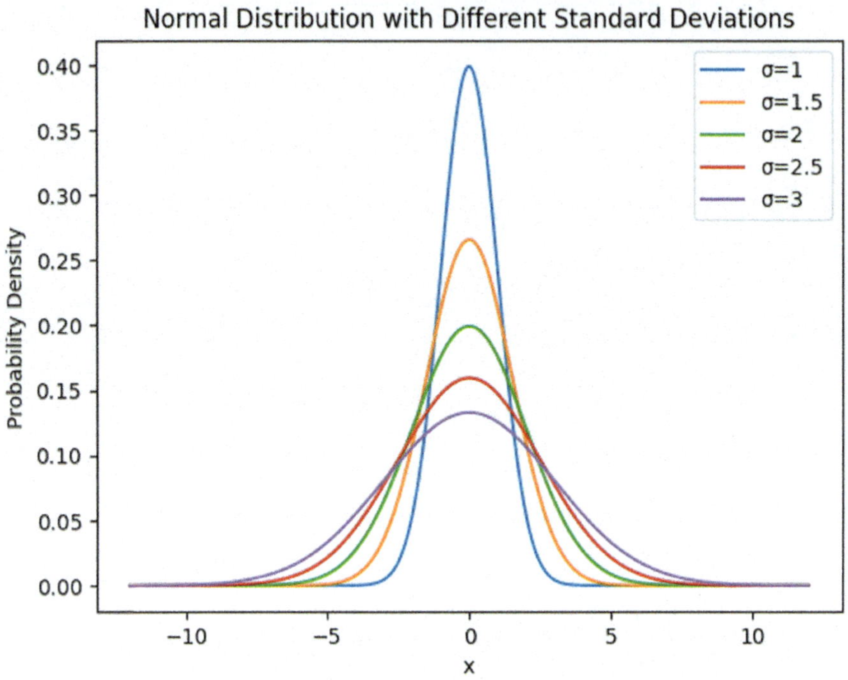


**Fig. 1.12** Normal distribution plot for different standard deviations

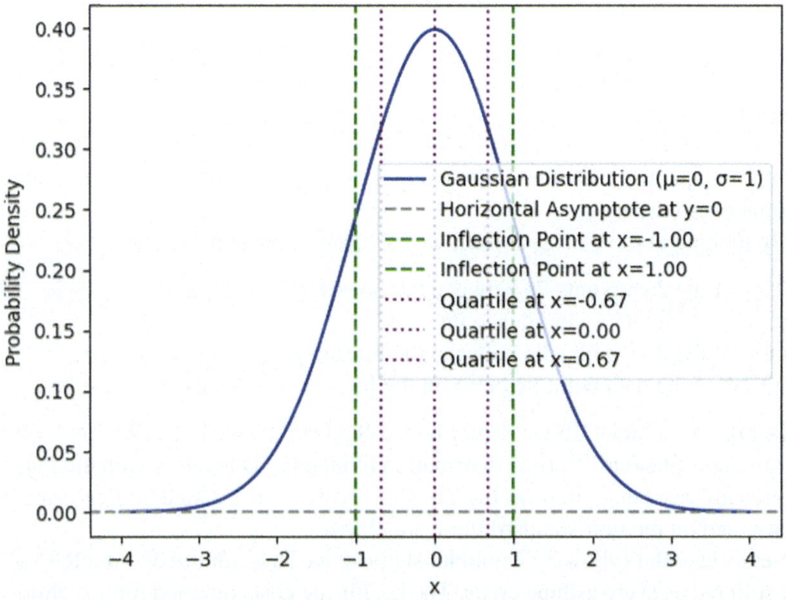


**Fig. 1.13** Gaussian distribution with asymptotes, inflection points, and quartiles

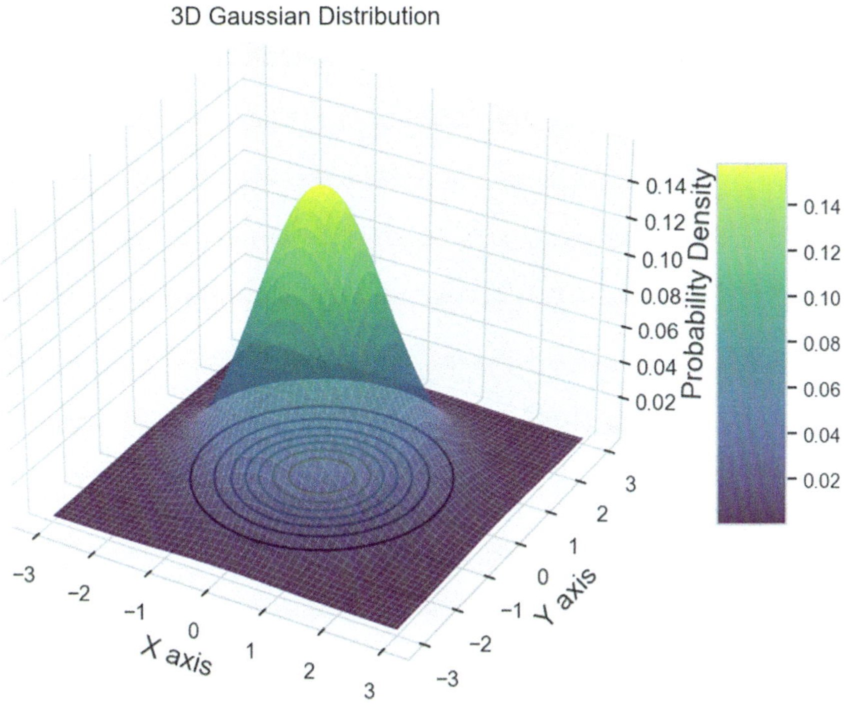


**Fig. 1.14** 3D standard normal distribution

major asthma crises will cost less than 6200 €, knowing that the average cost is 5912 € with a standard deviation of 200 €?

We shall solve this problem using the *standardized Z-score*. We will compute $Z$, using the following formula:

$$Z = \frac{x - \mu}{\sigma}$$

That makes $Z = \frac{6200-5912}{200} = 1.44$. We need to find the probability distribution associated with $Z = 1.44$. To find the corresponding probability we search for 1.44 in the corresponding table, A1 from the Appendix, and find the value 0.9251. Hence, approximately the costs for admitting in the ICU children that have suffered an asthma crisis will be in 93% of the cases lower than 6200 €.

***The Lognormal or Galton Distribution***

In Fig. 1.16 we have plotted the lognormal graph for different mean and standard deviation values.

When the growth rate of a certain process is independent of size, then we are dealing with a normally distributed logarithm, hence a lognormal distribution. In general, these types of distributions are right skewed, with a low mean and a large

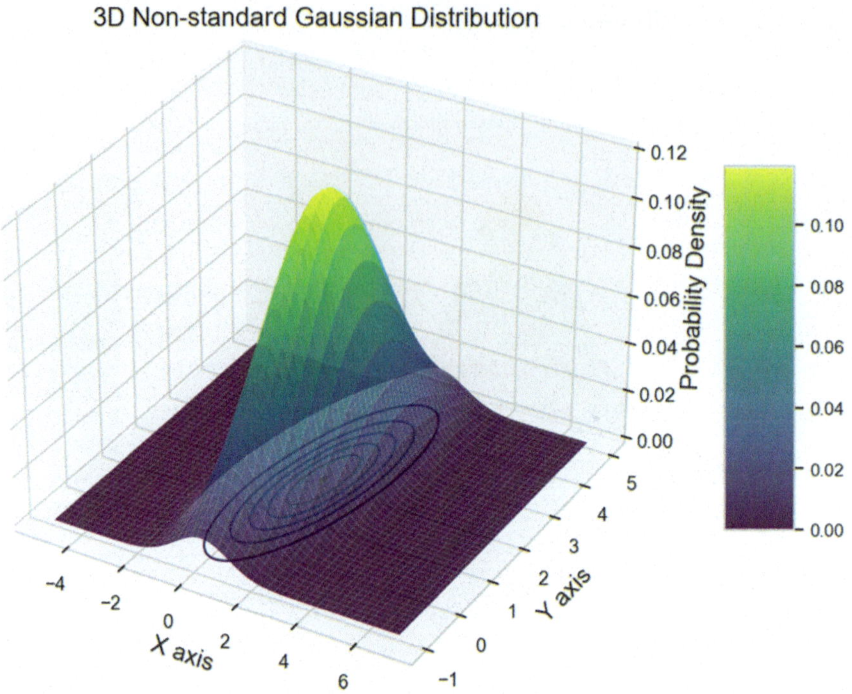


**Fig. 1.15** 3D Non-standard normal distribution

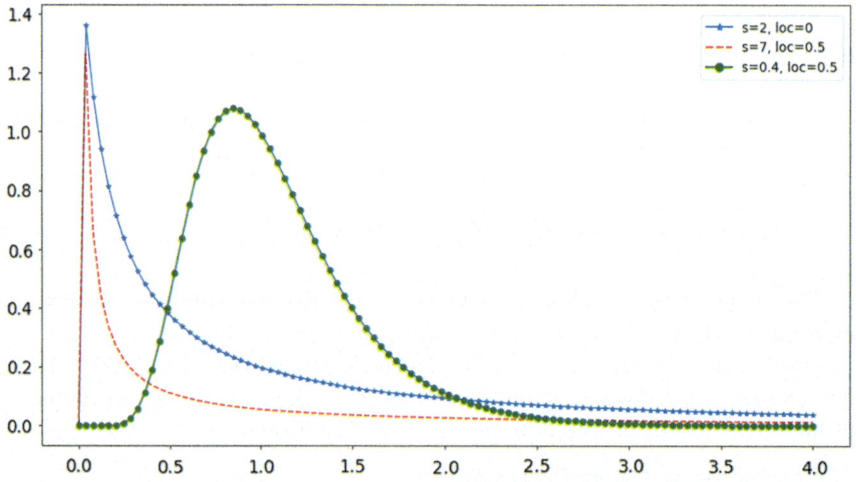


**Fig. 1.16** Lognormal distribution graphs for different mean and standard deviation parameters

variance. Since the logarithm $\log(x)$ exists, only if $x$ is positive, then all the data from the sample must be positive. If we need to transform a right skewed dataset into a dataset that is governed by the Normal distribution, we can use logarithms, [34–37]. The probability density function is defined by:

$$N(\ln x, \mu, \sigma) = \frac{1}{\sigma\sqrt{2\pi}} exp\left[-\frac{(\ln x - \mu)^2}{2\sigma^2}\right], x > 0.$$

Before diving into more statistical concepts, let's see how these distributions were discovered.

After determining which type of data we are dealing with, we need to perform EDA. EDA makes use of descriptive statistics and graphical representations to draw conclusions. EDA can be used to:

- Discover patterns in data;
- Spot anomalies;
- Identify the most important features;
- Make assumptions that need to be tested;
- Adjust hyperparameters;
- Link the observed phenomenon with certain causes;
- Built architectures for simple AI models.

***Descriptive Statistics***

You cannot use AI without Statistics. Actually, you can, but the results would be disastrous. People talk about trustworthy AI, and trustworthy AI without Statistics is impossible. Statistics is divided into two parts: descriptive and inferential. In this part of the book, we will discuss descriptive statistics. As the name says, we will describe the data using numerical parameters and graphical representations. AI works better when using large amounts of data. The more data we have, the more trustworthy and robust the results are. In a dataset we have objects, and those objects have features. The features take values from specific sets which are known as *statistical series*. The descriptive statistics is applied on the statistical series.

An object is seen from a computational point of view as an $n$-dimensional array, and from a probabilistic point of view as multidimentional *random variable*, each element being a random variable. Therefore, we can say that an object's feature $x$ is characterized by a random variable $X$. A *random variable* is a measurable function on the sample space$(\Omega, \Sigma)$. Its roots are from Probability Theory. On the other hand, a *statistical variable* indicates a measurable feature. Statistical variables come from Statistics. To complicate things even more, we denote both the random variable and statistical variable with $X$. Here is a trick to learn how to distinguish between the two concepts: when we discuss real objects from a dataset, we are dealing with statistical variables, so practical concepts; when we discuss abstract notions then we are dealing with random variables. Each random variable $X$ has a distribution function defined as:

$$F_X(x) = P\{X < x\} = P\{\omega \in \Omega; X(\omega)\epsilon(-\infty, x)\},$$

where $(\Omega, \Sigma, P)$ is the probability space.

The concept of *quantile* was introduced by Kendall in 1940. When we divide a sample into multiple adjacent subgroups of equal size we obtain quantiles. Depending on the number of quantiles we have *quartiles* (4 equal subgroups) or *percentiles* (100 equal subgroups), [38]. The quartiles are divided by three values $Q_1$, $Q_2$ and$Q_3$. 25% of the data lies below$Q_1$, which is the lower quartile or the 25th percentile, 50% of the data lies below $Q_2$, the 50th percentile, and 75% of the data lies below$Q_3$, the upper quartile or the 75th percentile. We can compare two distributions by plotting the Quantile–Quantile or Q-Q plot. Figure 1.17 presents a hypothetical Q-Q plot.

We can describe data using numerical parameters that show the central tendency, the distribution shape, and the data's variability. The central tendency can be determined by computing the *mean*, the *median*, or the *mode*.

The *mean* (*average* $\overline{\overline{x}}$) is the computed as:

$$\overline{x} = \frac{1}{n}\sum_{i=1}^{n} x_i,$$

in other words, the sum of all observations over the total number of observations. The data in the statistical series will be spanned around the mean's value.

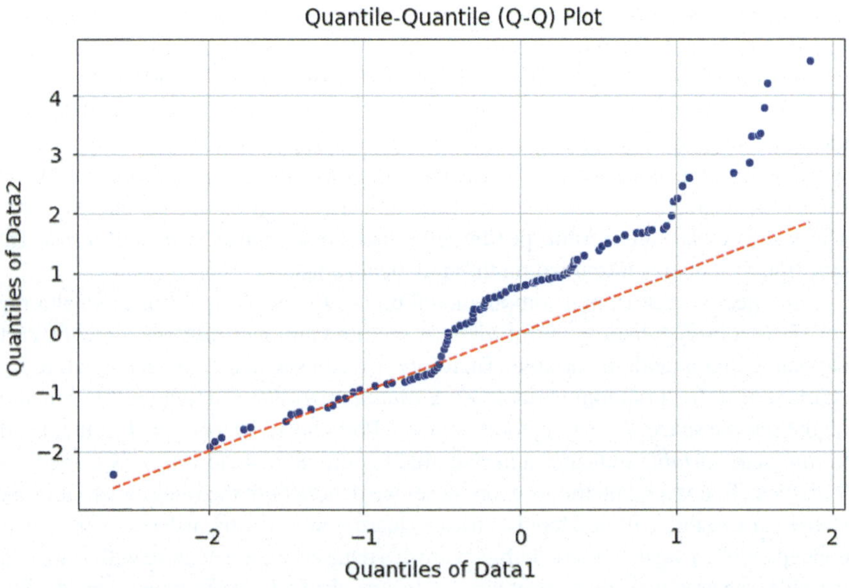


**Fig. 1.17** Q–Q plot for comparing distributions

For example, we have recorded the cost per day for 10 COVID-19 patients during the beginning of the pandemic in 2020:

$$\{2025, 1985, 2100, 1997, 2025, 1789, 2210, 2001, 1993, 2050\}.$$

The mean is computed as:

$$\begin{aligned} mean &= \frac{2025 + 1985 + 2100 + 1997 + 2032 + 1789 + 2210 + 2001 + 1993 + 2050}{10} \\ &= 2017.5 \end{aligned}$$

If our data does not have extreme values, we can use the mean parameter to compute the central tendency. However, if our data is sensitive to outliers, then we will use the *median* to compute the central tendency. When calculating the median, we arrange the data ascending or descending, and the median will be the second quartile, $Q_2$.

$$P\{X \leq Q_2\} = P\{X > Q_2\} = \frac{1}{2}.$$

If the sample size is even, then the median is the mean of the two middle observations, otherwise the median is the value of the middle observation. The median for the cost per day of COVID-19 patients is: 2013.

Figure 1.18 presents the mean versus the median in a data sample that has outliers.

The most frequently encountered value in a data sample is called the *mode*. A data sample that has only one value that is repeated a maximum number of times is called *uni-modal*, if it has two values that have the same maximum number of occurrences it is called *bi-modal*, and more than two values it is called *multi-modal*. The data from our example is uni-modal since we have two occurrences of 2025.

If the data is Normally distributed, then the mean, the median, and the mode are equal.

Another fact that we are interested in is how the data is spread around the mean. There are several parameters that help us determine this. For instance, we can compute the overall distance between every data sample and the mean. This is the *mean absolute deviation* (MAD) also known as the *absolute deviation from the mean*. We compute MAD as follows:

$$MAD = \frac{1}{n}\sum_{i=1}^{n}|x_i - \overline{x}|.$$

In the seventeenth century, Galileo Galilei was studying the stars. While he was making different astronomical measurements, he observed that the errors that occurred due to the imperfection of either the instruments or computations were

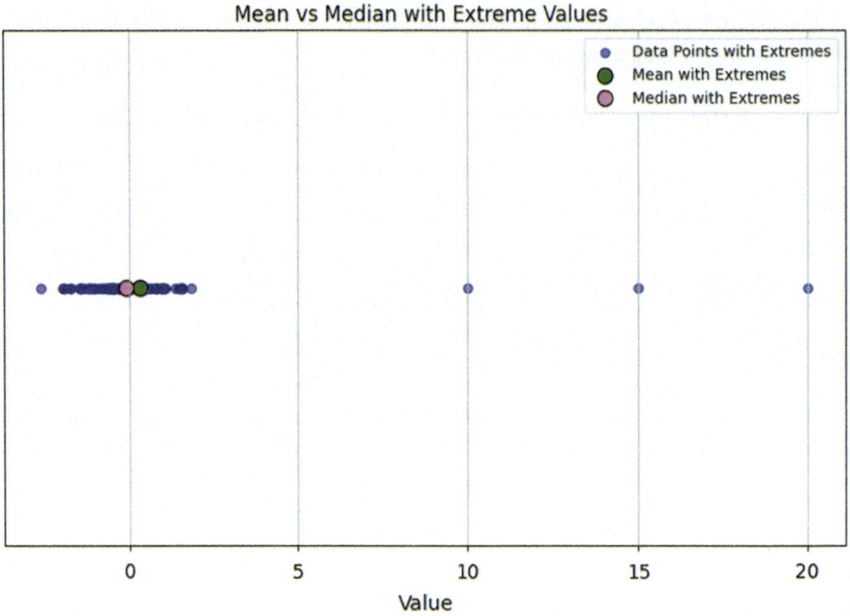


**Fig. 1.18** Mean versus median in population with outliers

symmetric. Another thing he noted was the fact that the smaller errors seem to occur more than large ones. Multiple assumptions were made regarding the distributions of these errors, but it was not until the early nineteenth century when mathematicians Adrain and Gauss, independently from each other, developed a formula for the Normal distribution and fit the errors on it. An interesting, yet not very known fact, is that the Normal distribution was discovered by Laplace in 1778, while developing the *central limit theorem*. Recall that in the Normal distribution, besides the mean, we have another important parameter, the *standard deviation*. The standard deviation is the root of the *variance*. In 1918, Fisher introduced in the Probability Theory, the concept of variance of a statistical series$\{x_i\}$, $i = 1, 2, \ldots, n$, as:

$$\sigma^2 = \frac{1}{n-1} \sum_{i=1}^{n} (x_i - \overline{x})^2.$$

Therefore, the formula for computing the standard deviation is:

$$\sigma = \sqrt{\frac{1}{n-1} \sum_{i=1}^{n} (x_i - \overline{x})^2}.$$

If the distribution is symmetric, then 95% of the data sample are found in the *confidence interval*, which is computed as *mean* $\pm\, 2 \cdot$ *standard deviation*.

So far, we have mentioned several times the word symmetric. How do you determine whether a distribution is symmetric or not? Besides by plotting it and observing the resulting graph? We can measure the asymmetry of a distribution by computing the *skewness*:

$$skewness = \frac{n \cdot \sum_{i=1}^{n} (x_i - \overline{x})^3}{(n-1) \cdot (n-2) \cdot \sigma^3}.$$

The obtained value can give us insights regarding whether the distribution is symmetric, or negatively skewed, or positively skewed. In Figs. 1.19, 1.20, and 1.21 we depict a symmetric, a negative and a positive skewed distribution.

We can describe a distribution using *kurtosis.* Kurtosis tells us if the data has few or many outliers, whether it is light or heavy tailed. There are three types of kurtosis: *metakurtic* (the value of the kurtosis is 0, and the distribution is comparable to the Normal one), *platykurtic* (the kurtosis is negative, the tail is light, therefore the data does not have many outliers), and *leptokurtic* (the kurtosis is positive, the tail is heavy, therefore the data has many outliers).

In Fig. 1.22 we present the above-described types of kurtoses.

The formula for computing the kurtosis is:

$$kurtosis = \frac{n \cdot (n+1) \cdot \sum_{i=1}^{n} (x_i - \overline{x})^4 - 3 \cdot (n-1)\left[\sum_{i=1}^{n} (x_i - \overline{x})^2\right]^2}{(n-1) \cdot (n-2) \cdot (n-3) \cdot \sigma^4}.$$

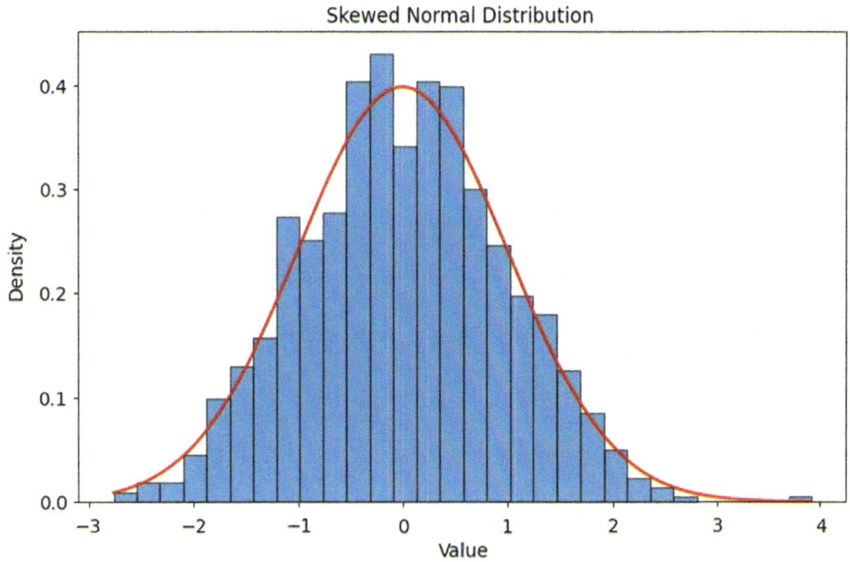


**Fig. 1.19** Symmetric distribution

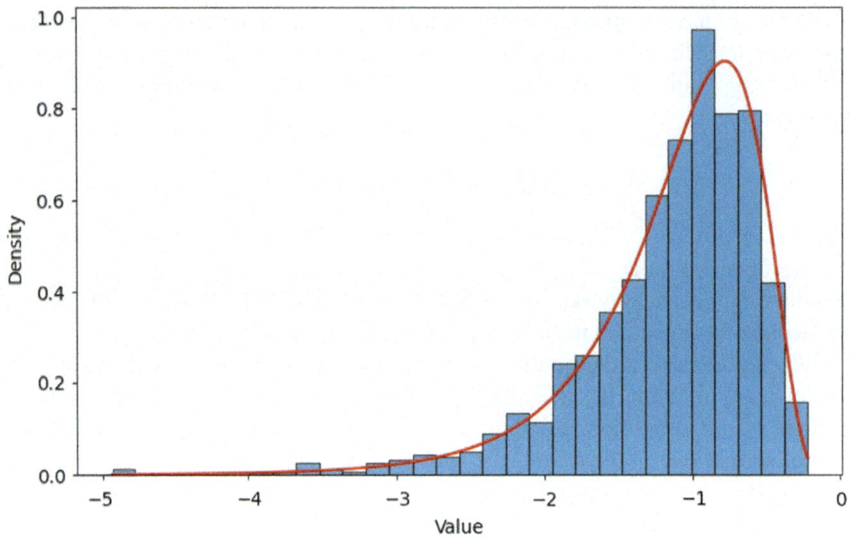


**Fig. 1.20** Negatively skewed distribution

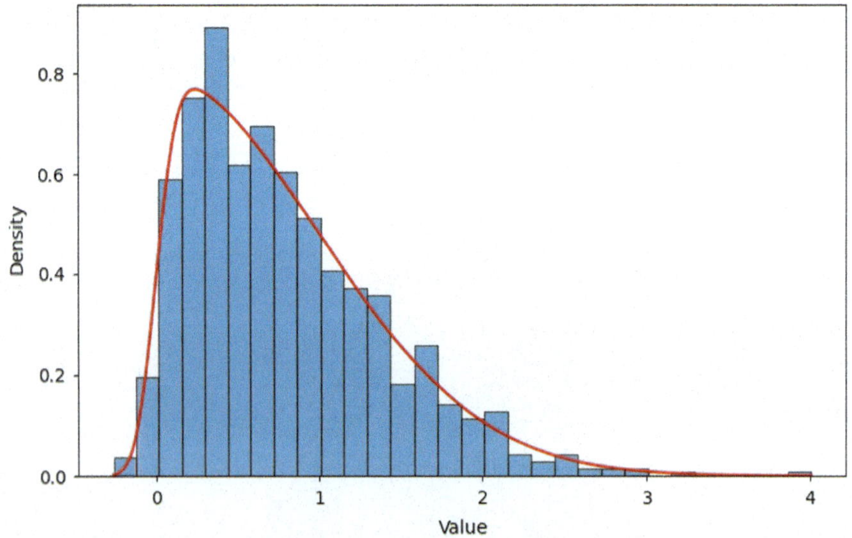


**Fig. 1.21** Positively skewed distribution

So far, we have discussed how we can describe one statistical variable using descriptive statistics. Let us see now what statistical concepts we can use to describe the relationship between two or more statistical variables. To describe the relationship between two statistical variables $(X, Y)$ that correspond to the couples of

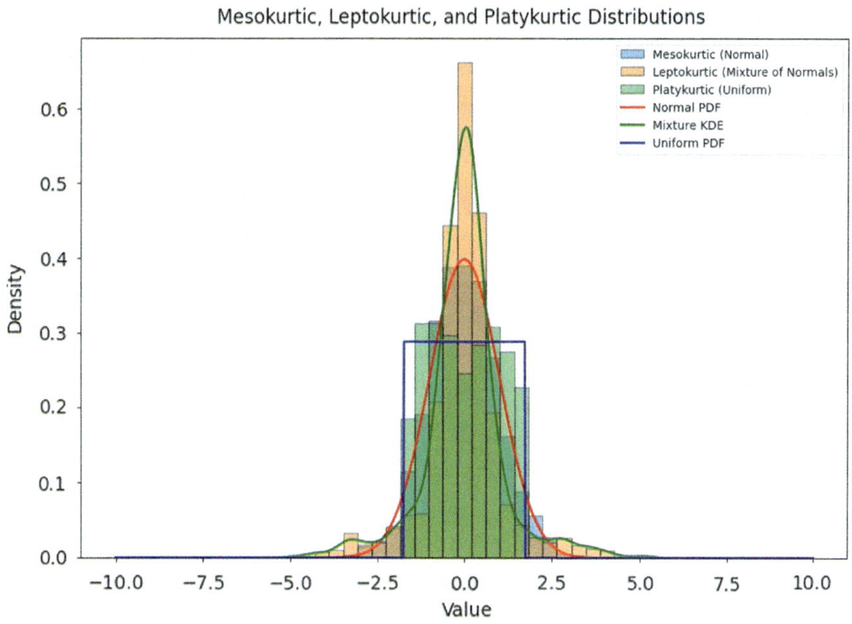


**Fig. 1.22** Mesokurtic, leptokurtic, and platykurtic graphs

values$(x_i, y_i)$. we can use the *correlation coefficient,r*. Please keep in mind that in order to compute it the values must refer the same object. The value of the correlation coefficient can indicate whether the variables are not linearly correlated ($r \sim 0$), whether they are negatively correlated ($r \sim -1$), that is when the value of the first goes up, the value of the other goes down, or whether they are positively correlated ($r \sim +1$), that is when the value of the first goes up, the value of the second goes up also.

The correlation coefficient is computed by applying a mathematical formula on two variables. For our result to be statistically relevant, we first need to verify whether the following two assumptions are satisfied or not:

1. The two variables are defined on the same dataset and refer to the same object.
2. The Normal distribution governs at least one variable. If this is not the case, we have two possibilities: either we use a non-parametric correlation coefficient like the rank correlation, or we transform the data, so it has an approximately Normal distribution.

To measure the correlation between two statistical variables $(X, Y)$, we need to compute the *covariance* between the two corresponding statistical series $\{x_i\}, i = 1, 2 \ldots, n$, and $\{y_i\}, i = 1, 2, \ldots, n$, as follows:

$$cov(X, Y) = \frac{1}{n} \sum_{i=1}^{n} x_i y_i - \overline{xy}.$$

Just like in the case of discovering the Normal distribution, when three different persons independently had the same idea, the *correlation* coefficient was discovered by a French physicist August Bravais, and by the English statistician Sir Francis Galton, when both have observed that there is a link between a person's height and the length of his/her forearm, [39, 40]. Galton's colleague, Pearson, extended Galton's idea from simple linear regression to multiple linear regression, and named the correlation coefficient as *Pearson's r*. Its formula is:

$$r = \frac{\sum_{i=1}^{n}(x_i - \overline{x}) \cdot (y_i - \overline{y})}{\sqrt{\sum_{i=1}^{n}(x_i - \overline{x})^2 \cdot \sum_{i=1}^{n}(y_i - \overline{y})^2}}.$$

The above equation can be rewritten under the following form:

$$r = \frac{\sum_{i=1}^{n}x_i y_i - \frac{1}{n}\sum_{i=1}^{n}x_i\sum_{i=1}^{n}y_i}{\sqrt{\left[\sum_{i=1}^{n}x_i^2 - \frac{1}{n}\left(\sum_{i=1}^{n}x_i\right)^2\right]\left[\sum_{i=1}^{n}y_i^2 - \frac{1}{n}\left(\sum_{i=1}^{n}y_i\right)^2\right]}}.$$

Depending on $r$'s value we can find out whether the variables are correlated or not. Its value domain is $[-1, 1]$. As $r$'s value approaches $-1$ or 1, we know that the variables are strongly correlated, either negatively, or positively. If it is a negative correlation, then as one value increases the other one decreases, if it is positive correlation, then both variables increase, or both decrease. As $r$'s value reaches 0, the correlation between the variables weakens.

If two variables are plotted as a scatter diagram, and the 'cloud' or points has an ellipse shape, then we can state that the variables are correlated. $r$ has as correspondence the elongation of this ellipse.

From Figs. 1.23 and 1.24 we can clearly see the differences between the graphs of correlated and uncorrelated variables.

Besides using the descriptive statistical parameters to understand data, we can use graphical representations to provide insights. In the following section we shall present different graphs and their interpretations.

For example, we are going to presume that we are the manager of a hospital that has 12 departments: cardiology, neurology, oncology, pediatrics, orthopedics, neonatology, ICU, surgical, burn unit, maternity, trauma, and emergency care. In Fig. 1.25, we have listed the number of beds in each department, the number of doctors and nurses, the cost per day of patient, and whether the department is profitable or not.

We are interested in seeing the connection between different variables plotted. Figures 1.26, 1.27, 1.28, 1.29, 1.30, 1.31, 1.32, 1.33, 1.34, 1.35 and 1.36 have been obtained using *seaborn* package from Python. Figure 1.26 represents a scatter plot, where on the $X$-axis we have the number of beds, and on the $Y$-axis, we have the number of nurses. In the graph we have two kinds of bullets, the green one means that the department is profitable, the red one that it is not. From the scatter plot we can see that the variables number of beds and number of nurses are linearly dependent.

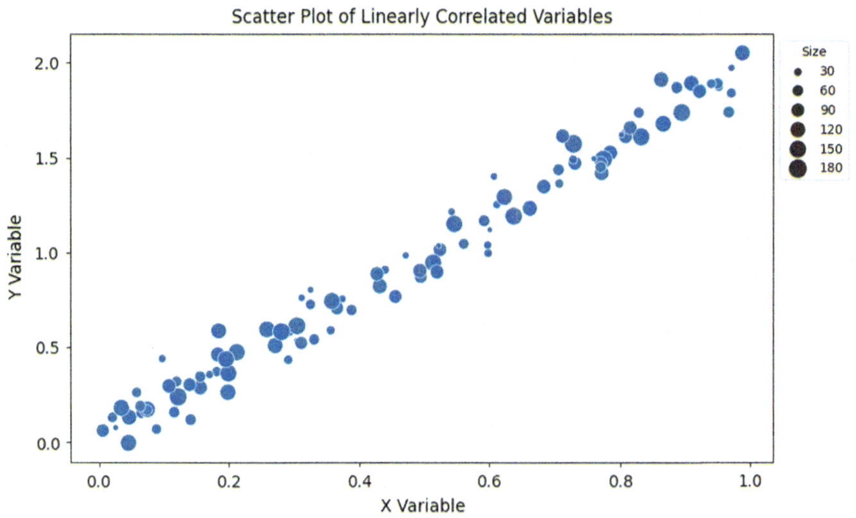


**Fig. 1.23** Scatter plot of correlated variables

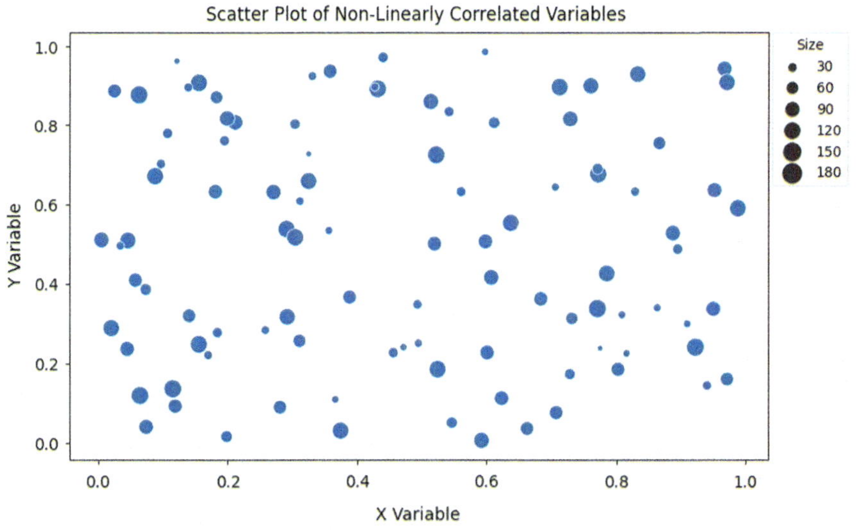


**Fig. 1.24** Scatter plot of uncorrelated variables

To illustrate other types of graphs, we will use another example. In Fig. 1.27 we describe 11 premature patients admitted to the Neonatology ward. For instance, Patient_1 was born at 29 weeks, weighted 1871 g, had an Apgar score of 5, the severity of illness was moderate, spent 25 days in hospital, the overall cost was 20,609 €, and was admitted to the NICU. Please take note that this data is artificially generated to illustrate potential graphs. Some inconsistencies might appear.

| | department | no_beds | staff_doctors | staff_nurses | cost_per_day | profitable |
|---|---|---|---|---|---|---|
| **0** | Cardiology | 50 | 15 | 30 | 1200 | True |
| **1** | Neurology | 40 | 12 | 25 | 1500 | True |
| **2** | Oncology | 30 | 10 | 20 | 1800 | False |
| **3** | Pediatrics | 60 | 20 | 40 | 800 | True |
| **4** | Orthopedics | 45 | 14 | 28 | 1300 | False |
| **5** | Neonatology | 20 | 8 | 15 | 2000 | True |
| **6** | ICU | 25 | 10 | 20 | 2500 | True |
| **7** | Surgical | 35 | 18 | 25 | 1700 | True |
| **8** | Burn Unit | 10 | 5 | 12 | 2200 | False |
| **9** | Maternity | 50 | 12 | 35 | 900 | True |
| **10** | Trauma | 15 | 9 | 18 | 1600 | True |
| **11** | Emergency Care | 20 | 11 | 22 | 1400 | True |

**Fig. 1.25** Data from an imaginary hospital

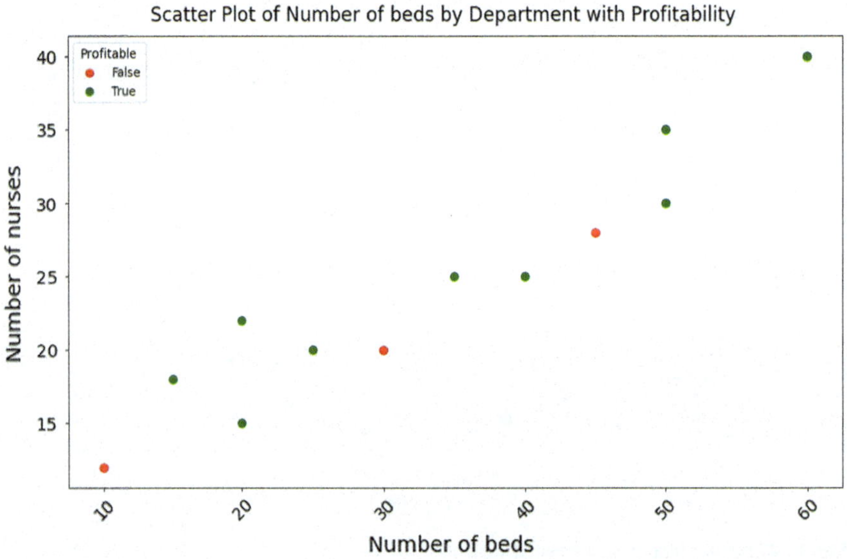


**Fig. 1.26** Scatter plot using the variables number of beds and number of nurses

| | Patient | Age_in_Weeks_at_Birth | Apgar_Score | Birth_Weight_in_Grams | Severity_of_Illness | Days_Spent_in_Hospital | Total_Cost | Stay_in_NICU |
|---|---|---|---|---|---|---|---|---|
| 0 | 1 | 29 | 5 | 1871.460590 | moderate | 25 | 20608.625525 | True |
| 1 | 2 | 26 | 4 | 1122.087955 | moderate | 53 | 74561.907402 | True |
| 2 | 3 | 35 | 9 | 2326.558699 | low | 29 | 38222.428212 | False |
| 3 | 4 | 37 | 9 | 3032.876700 | low | 10 | 10107.473026 | False |
| 4 | 5 | 33 | 7 | 2500.893204 | mild | 5 | 3099.326837 | False |
| 5 | 6 | 30 | 6 | 1337.615171 | mild | 46 | 46864.368600 | True |
| 6 | 7 | 35 | 7 | 2843.622751 | mild | 21 | 15832.223693 | False |
| 7 | 8 | 27 | 6 | 1246.876063 | mild | 16 | 19396.329439 | True |
| 8 | 9 | 29 | 4 | 1148.086930 | moderate | 16 | 16042.864372 | True |
| 9 | 10 | 32 | 7 | 2061.774501 | mild | 5 | 5671.756724 | True |
| 10 | 11 | 25 | 1 | 840.352726 | severe | 73 | 69188.171014 | True |

**Fig. 1.27** Snapshot of the first 10 artificially generated premature patients

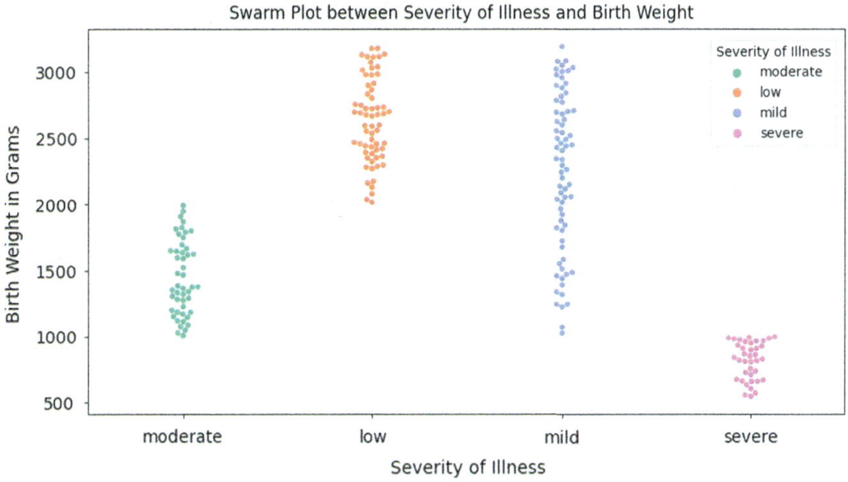


**Fig. 1.28** Swarm plot for the birth weight and severity of illness

Figure 1.28 presents a swarm plot regarding the birth weight and the severity of illness. Technically, the swarm plot stacks all similar points together. We have used different colors for the severity for a better visualization of the situation. Figures 1.29 and 1.30 show the correlation matrix in the form of a heatmap. The first image, 1.29 shows only the colors, without numbers. As the correlation strength increases or decreases, the colors change in hue. A strong positive correlation is represented with a dark color, whereas a strong negative correlation with a light color. Without numbers attached to colors, we can't make much of this heatmap. Figure 1.30 shows the correlation heatmap together with the *r* coefficient.

Figure 1.31 shows a box-and-whiskers plot for the age in weeks at birth.

The line inside the rectangle represents the mean, the rectangle represents the mean + standard deviations. The 95% confidence interval is represented by the whiskers (the lines outside the rectangle). If we look at Fig. 1.32, in which we have two box-and-whiskers plots regarding the birth weight and the stay in the NICU, we

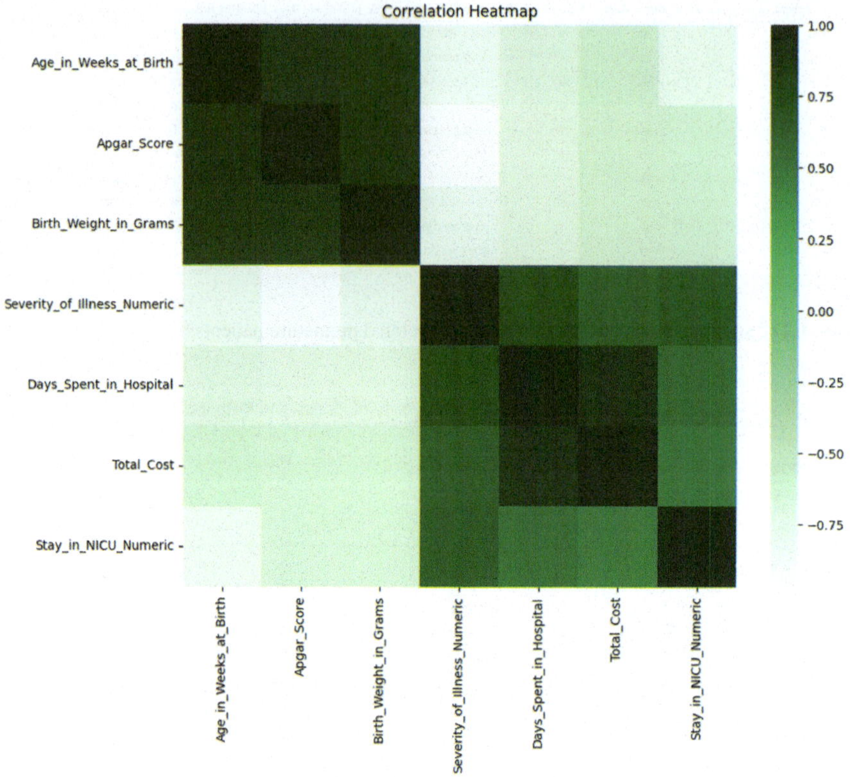


**Fig. 1.29** Correlation matrix heatmap without the correlation coefficient

see that there are dots inside the plots. The dots represent the data points. The points that are above or below the confidence interval represent the outliers.

Figure 1.33 depicts the distribution of birth weights and admission to the NICU. We can see that most patients that had a low birth weight were admitted to the NICU.

In Fig. 1.34 we present the bar plot that corresponds to the Apgar score. In this scenario most of our patients have the Apgar score equal to 7. Recall that this is an artificially generated example, hence this plot does not reflect the general real population.

The following two graphs (Figs. 1.35. and 1.36) plot the density to reflect the distribution between the total costs and the days spent in hospital, or the density of the distributions between the severity of illness and the Apgar score.

Visualizing the data helps the hospital manager have a clear picture of the situation, and act accordingly to the best interest of the patients, as well as for the financial status of the facility.

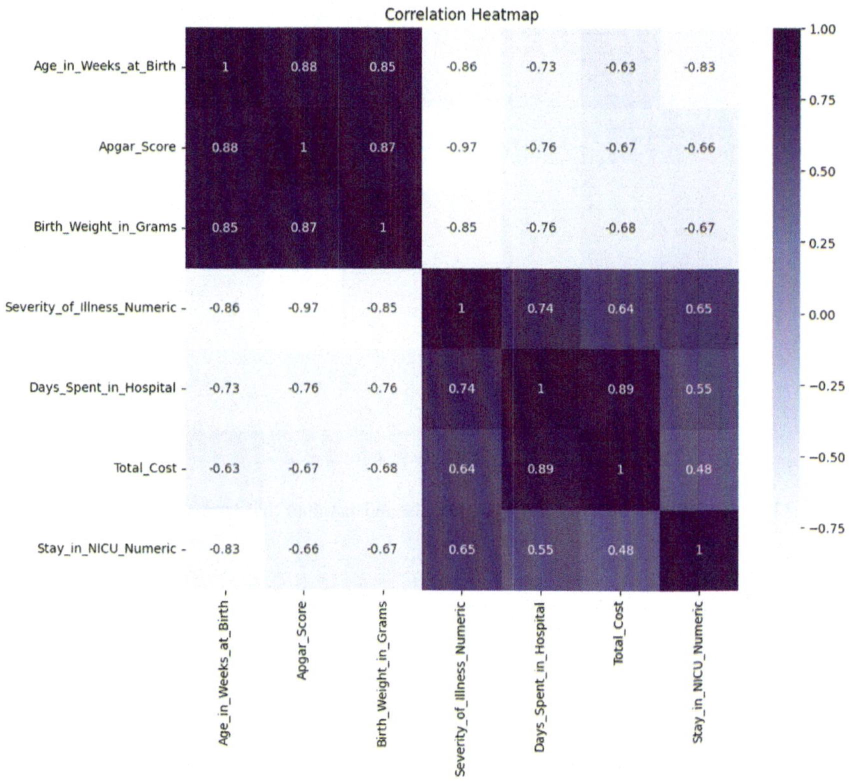


**Fig. 1.30** Correlation matrix heatmap with correlation coefficient attached

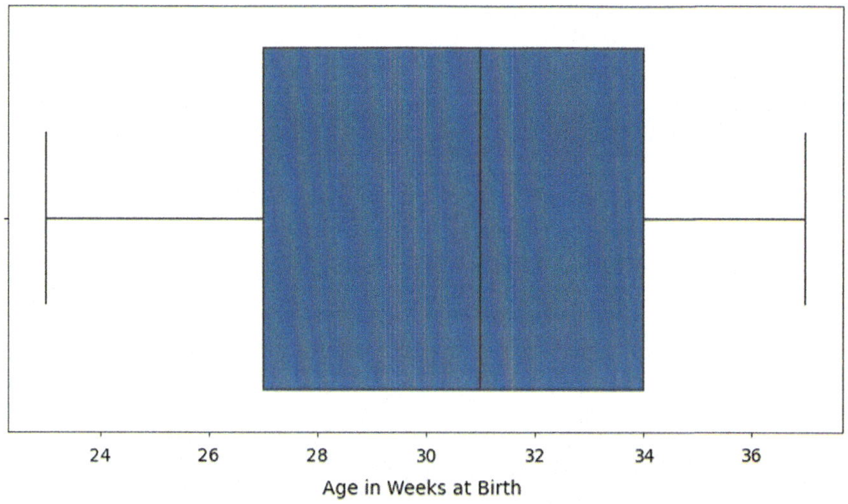


**Fig. 1.31** Box-and-whiskers plot for the age in weeks at birth

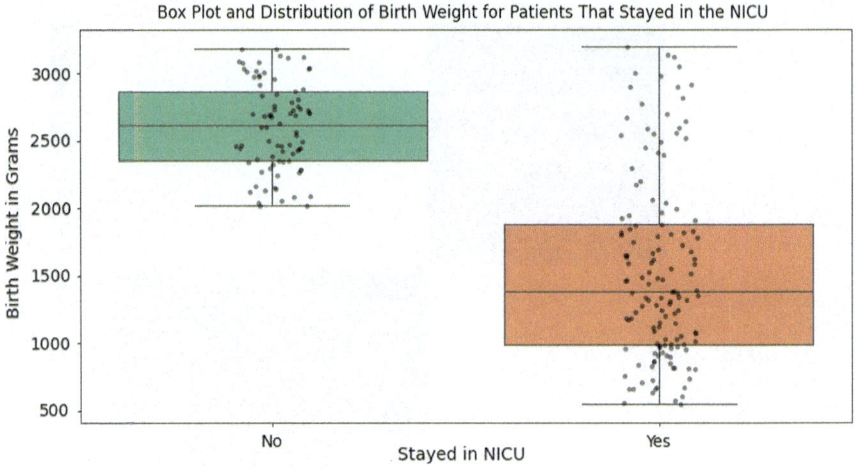


**Fig. 1.32** Box-and-whiskers plots of birth weight and patients that were admitted or not in the NICU

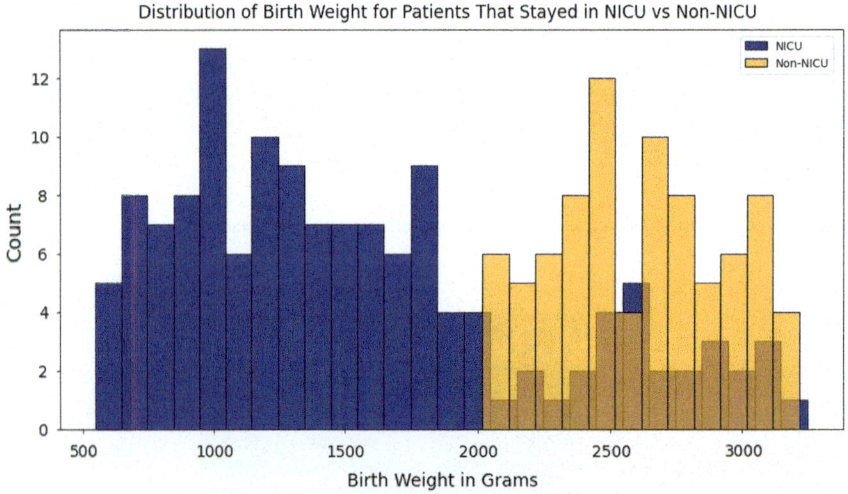


**Fig. 1.33** Distribution of birth weights and whether the patients were admitted to the NICU

The last part of EDA that regards hypothesis testing will be covered in the next chapter.

Having knowledge regarding quantitative data helps us perform quantitative analysis, which is ultimately a team effort. A team of experts must transform a general situation into a well-defined mathematical problem that can be further on approached

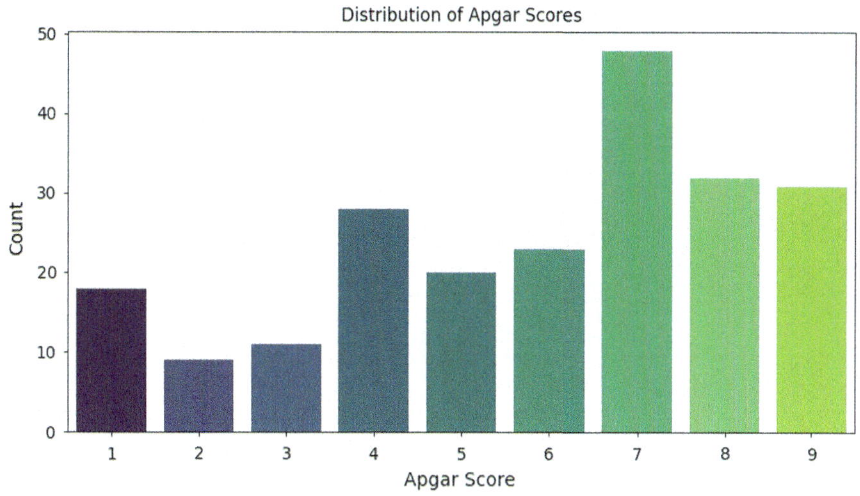


**Fig. 1.34** Bar plot regarding the Apgar score

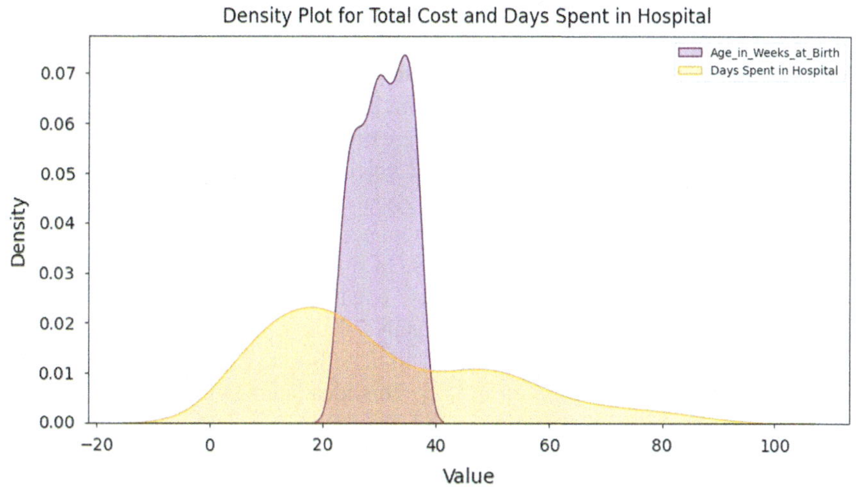


**Fig. 1.35** Density plot for total cost and days spent in hospital

quantitatively. The difference between failure and success is made by the hospital manager. Working together with the data scientist and understanding every step of the process, will lead to success, otherwise to failure. After the data scientist and hospital manager have a discussion regarding the situation that needs to be handled, she/he models the problem and presents various candidate solutions to the manager. It is the manager's responsibility to choose the "best" one.

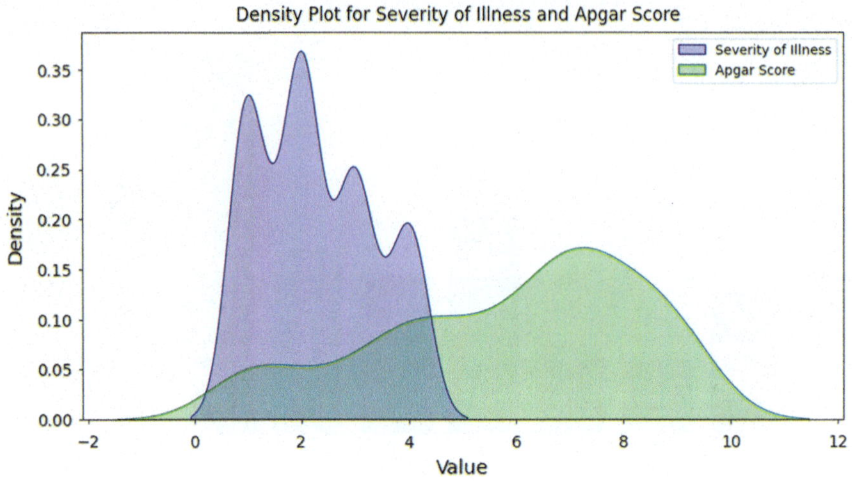


**Fig. 1.36** Density plot for the severity of illness and Apgar Score

## 1.3 Cost, Revenue, and Profit. How to Model Them?

Any situation or object can be represented through a *model*. Models can be *iconic, analog*, or *mathematical*, [41]. An *iconic* model is a scale representation of something. The best examples for iconic models are the Playmobil or Lego toys. They are realistic physical smaller replicas of actual objects and model different scenarios. The *analog* model is not an identical physical replica, but still mimics a physical phenomenon. For example, the thermometer, speedometer, or blood pressure measuring device represent analog models. In this book, we will cover the *mathematical* models. Mathematical models use equations and symbols to model any problem. They are critical in quantitative analysis, and decision-making.

Data scientists will model a problem, simulate it, and give potential solutions, without actually experimenting on it, [42]. This idea belongs to Stanislaw Ulam, a Polish American mathematician, and John Von Neumann, a Hungarian American mathematician. During World War II, the two mathematicians and other brilliant scientists worked on the Manhattan Project to build the Atomic Bomb. Using Von Neumann's ENIAC (Eletronic Numerical Integrator and Computer), they tried to compute Ulam's Monte Carlo calculations. It is obvious that they had to model and simulate the Atomic Bomb to see if it works, without actually experimenting it.

What was the actual situation: in 1930s, the idea that an atom can be divided through nuclear fusion appeared. If we divide only one atom, the released power doesn't do anything. A chain reaction, on the other hand, works like domino, and releases the energy of one atom after another, leading to a nuclear bomb. This was the theory. In practice, three possibilities arose: (1) the chain reaction is non-existent; (2) The chain reaction happens, and the bomb explodes; and (3) the chain reaction happens but the bomb does not go off. Obviously, none of the creators of the nuclear bomb wanted to test it, and they prefer simulating the result.

This implied the mathematical modeling of the atom: how much energy it will release, what is the probability to explode, etc. Each atom behaves differently, so all these variables must be modeled through resolving an integral equation.

Besides the above-mentioned models, another model has been introduced by Brian Wilson, the *conceptual* model which adds qualitative information, [43].

Let us think of the following example: we are the hospital manager, and there has been a massive car crash. We are expected to receive a lot of casualties in the ER. The cost of the ambulance ride to our hospital is around \$900, while the ER visit is around \$3300. The average inpatient hospitalization following a car crash is around \$58,000. If we add these values, we obtain \$62,200 per patient. By denoting with $x$ the number of casualties, we can model the total cost of care as $C = 62{,}200 \times x$.

This model will enable us to make different scenarios and inferences about the real situation. So, if there are 4 casualties, then the total cost is \$248,800. This simple mathematical model is a quick way of identification of total cost expectations. The difference between a good and a bad decision is given by how well the model mimics the real-life scenario. The more closely the mathematical model resembles real life, the more accurate the predictions will be.

In general, when we deal with a hospital managerial problem, we find ourselves faced with an optimization problem, whether we need to maximize the profit or minimize the cost. Obviously, we have constraints and restrictions such as the number of beds, staff, operating rooms, drugs, etc. For a successful decision, we need to model carefully the objective and the constraints in the form of mathematical equations. The problem objective represents the *objective function*. If we were to consider the above equation that computes the total cost of a car crash injury, the objective function would be to maximize the number of patients we treat, so that the insurance companies pay the hospital more money. Let us add a restriction: to give proper care in the trauma room, the doctors need at least 1 h, and we have only 5 E.R. doctors that work in 12 h shifts. For each patient we need at least 2 doctors. Hence, in one hour 2 patients can be seen, and in a shift 24 patients. But these are not the only constraints. You just cannot put on hold trauma patients, and you cannot close the ER for other incoming cases. The decision problem would be: how many patients can your hospital take, and how many should you turn away. Will you close the trauma center for other incoming cases?

In every mathematical model we have parameters that we can control such as the cost per patient, and the time to patch up each patient, and parameters that are out of control: such as how many patients will come, and how complicated are the cases, etc. These parameters are known under the name of *uncontrollable* and *controllable inputs*. The controllable inputs represent the *decision variables* of the model. The moment we have all controllable and uncontrollable inputs specified, we are able to evaluate the objective function, and determine the model's output. We can play with different inputs and make up what-if scenarios to project particular factors that might occur in a real situation. In Fig. 1.37 we present a flowchart of how you project controllable and uncontrollable variables in a mathematical model.

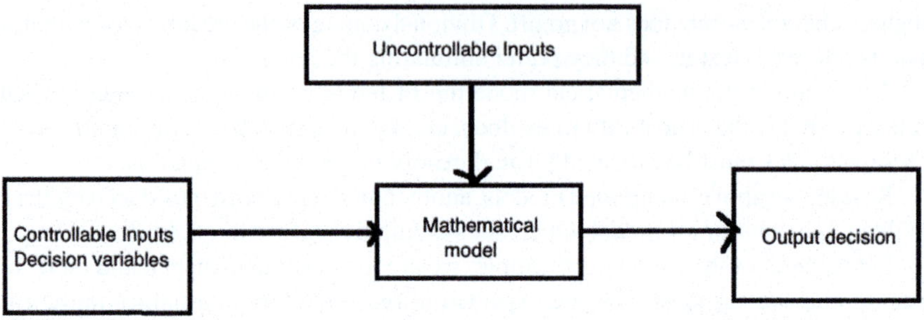


**Fig. 1.37** Flowchart of mathematical modelling

In our example the number of patients that can be seen in a shift is an uncontrollable input, because the hospital manager cannot control how many hours are in a shift. However, the manager can hire more doctors, transforming thus an uncontrollable parameter into a controllable one.

The uncontrollable inputs can be constant, or they can vary. If they are constant, then our model is a *deterministic* model. The number of hours in a shift is such a parameter. If the number of hours in a shift is the only uncontrollable input in a model, then that model would be a deterministic model. Another aspect that we need to specify is that in a deterministic model, the uncontrolled input values are known a priori.

If the uncontrollable inputs vary, then our model is *stochastic* or *probabilistic*. In our example, the number of hours needed to take care of a patient varies, hence making our model a stochastic one.

An important step in quantitative analysis is represented by data preparation or data engineering. The uncontrollable input is not instantly ready to be used. It takes a team effort to prepare them. We should not collect data as we develop the model. We should decide which data is needed, adopt a notation for it, and then perform a data preparation step to gather it.

One should not consider the data preparation step a trivial one. It is of high importance. Recall the saying *garbage-in garbage-out.* Errors in data gathering reflect on the model's performance. Another thing that we should consider is that the larger the dataset, the more accurate the result.

Once the data preparation step is over, and the model has been developed, we can proceed to optimize our output. Some procedures use the trial-and-error approach, but in this book, we shall use AI algorithms such swarm intelligence to optimize the objective functions. Multiple AI algorithms can be used to optimize the objective function. Our job is to determine the best suited one for every situation. Since we are discussing stochastic models, we are going to use statistical tests to determine the appropriate algorithm. More about statistical tests, what are they and how to use them in the next chapters.

After we obtain a model solution, the data scientist and the hospital manager will want to see how good the obtained solution really is. This step involves model testing and validation. If the model solution passes this step, then it has a green light to be used on a full-scale management situation. If the model solution fails this step, different remedial actions are used to mend the situation, such as modifying the mathematical model, using a different AI algorithm to optimize the objective function, or gather more accurate data for the training process. The solution model will never be put into practice until it passes the testing and validation steps.

In this chapter we have discussed a little bit of the hospitals' history, and some parts of quantitative analysis such as types of data, data distributions, and descriptive statistics. We have learned about different types of modeling and types of input parameters. In the next chapters, we shall continue some aspects of the quantitative analysis such as statistical tests, how to use and interpret their results, and also discuss data gathering and engineering.

## References


1. Bertman, S.: Handbook to Life in Ancient Mesopotamia. Oxford University Press (2003)
2. Kramer, S.N.: Everyday Life in Ancient Mesopotamia. University of Pennsylvania Press (1988)
3. Kramer, S.N.: The Sumerians: Their History, Culture, and Character. University of Chicago Press (1971)
4. Nemet-Nejat, K.R.: Daily Life in Ancient Mesopotamia. Greenwood Press, CT (1998)
5. Riva, M.A., Cesana, G.: The charity and the care: the origin and the evolution of hospitals. Eur. J. Int. Med. **24**(1), 1–4 (2013)
6. Aitken, J.T., Fuller, H.W.C., Johnson, D.: The Influence of Christians in Medicine. Nottingham Intervarsity Press (1984)
7. Haeger, K.: The Illustrated History of Medicine. London Harold Starke (1988)
8. Woodhead, A.G.: The state health service in ancient Greece. Camb. Hist. J. **10**, 235–253 (1952)
9. Scheider, K.: Valentudinarium, Realencyclopaedie der Altertumswissenschaft Bad, VII.A.1, pp. 262–326. Mittlerische Buchgesellschaft, Leipzig (1953)
10. Thompson, J.D., Golden, G.: The Hospital: A Social and Architectural Study. Yale University Press, New Have & London (1975)
11. Jackson, R.: Doctors and Diseases in the Roman Empire. University of Oklahoma Press, London (1988)
12. Staccioli, E.E.R.: Ancient Rome. Moments Past and Present. Rome, La Zincografica Fiorentina (1989)
13. Mundy, J.H.: The High Middle Ages. Logman Group UK Ltd., London (1998)
14. Major, R.H.: A History of Medicine. Springfield III. Charles C. Thomas (1954)
15. Porter, R.: The Greatest Benefit to Mankind. Harper Collins, London (1997)
16. Guthrie, D.: A History of Medicine. Thos. Nelson & Sons, Edinburgh (1958)
17. Cilliers, L., Retief, F.P.: The evolution of the hospital from antiquity to the end of the Middle Ages. Curations **25**(4), 60–66 (2002)
18. Stevens, S.S.: On the theory of scales of measurement. Sci. New. Ser. **103**(2684), 677–680 (1946)
19. Kirch, W.: Level of measurement. In: Encyclopedia of Public Health, vol. 2, pp. 851–852. Springer (2008). https://doi.org/10.1007/978-1-4020-5614-7_1971
20. Mitchell, J.: Measurement scales and statistics: a clash of paradigms. Psychol. Bull. **100**(3), 398–407 (1986). https://doi.org/10.1037/0033-2909.100.3.398
21. Papoulis, A.: Probability, Random Variables, and Stochastic Processes, 2nd edn, pp. 108–109. McGraw-Hill, New York (1992)
22. Spiegel, M.R.: Theory and Problems of Probability and Statistics. McGraw-Hill, (2003)

23. Steinhaus, H.: Mathematical Snapshots, 3rd edn. Dover, New York (1999)
24. Lindstrom, D.: Schaum's easy Outline of Statistics, 2nd edn. McGraw-Hill Education (2010)
25. Vogt, W.P.: Dictionary of Statistics and Methodology: A Nontechnical Guide for the Social Sicences. SAGE (2005)
26. Wheelan, C.: Naked Statistics. W.W. Norton & Company (2014)
27. Altman, D.G.: Practical Statistics for Medical Research. Chapman and Hall (1991)
28. Beyer, W.H.: CRC Standard Mathematical Tables, 28th edn., pp. 533–534. CRC Press, BocaRaton, FL (1987)
29. Feller, W.: An Introduction to Probability Theory and its Applications, vol. 1, 3rd edn. Wiley, New York (1968)
30. Kenney, J.F., Keeping, E.S.: Mathematics of Statistics, vol. 2, 2nd edn. Von Nostrand, Princeton, NJ (1951)
31. Kjos-Hanssen, B.: Statistics for Calculus Students (2019). https://dspace.lib.hawaii.edu/handle/10790/4572
32. Kraitchik, M.: The Error Curve, Mathematical Recreations, pp. 121–123. W.W. Norton, New York (1942)
33. Patel, J.K., Read, C.B.: Handbook of the Normal Distribution. Dekker, New York (1982)
34. Aitchison, J., Brown, J.A.: The Lognormal Distribution with Special Reference to its Use in Economics. Cambridge University Press, New York (1957)
35. Balakrishnan, N., Chen, W.W.S.: Handbook of Tables for Order Statistics from Lognormal Distributions with Applications. Kluwer, Amsterdam (1999)
36. Crow, E.L., Shimizu, K. (eds.): Lognormal Distributions: Theory and Applications. Dekker, New York (1988)
37. Limpert, E., Stahel, W.A., Abbt, M.: Lognormal distributions across the sciences: keys and clues: on the charms of statistics, and how mechanical models resembling gambling machines offer a ling to a handy way to characterize log-normal distributions, which can provide offer a ling to a handy way to characterize log-normal distributions, which can provide deeper insight into variability and probability-normal or log-normal: that is the question. Bioscience **51**(5), 341–352 (2001)
38. Galton, F.: Anthropometric percentiles. Nature **31**, 223–225 (1885)
39. Bravais, A.: Analyse mathematique sur le probabilities des erreus de situation d'un point. Memoires presents par divers savants a l'Academie des Sciences de France. Sci. Math. Phys. **9**, 255–332 (1846)
40. Galton, F.: Co-relations and their measurement chiefly from anthropometric data. Proc. R. Soc. Lond. **45**, 273–279 (1888)
41. Ackoff, R.L.: Scientific Method: Optimizing Applied Research Decisions. Wiley, New York (1962)
42. McClean, S.I.: Modelling and Simulation for Health Applications, Modeling Hospital Resource use. A Different Approach to the Planning and Control of Health Care Systems, pp. 21–27. Royal Society of Medicine Press (1994)
43. Wilson, R., Hancock, E., Bors, A., Smith, W. (eds.): Computer Analysis of Images and Patterns. Springer (2013)

# Chapter 2
# Clinical Department

**Abstract** The aim of this chapter is to present the manner in which the clinical departments are organized nowadays in hospitals. We will present different examples, from different countries, some developed, others underdeveloped. Our purpose is to see how we can make a difference in making the clinical department smarter, in each situation.

As medicine progresses, the healthcare systems are overwhelmed by different challenges such as: diagnostic and treatment costs that are escalating, limited access to medical help, and a growing demand for personalized care. These issues concern every country, no matter whether it is a developing or a developed one. A hospital manager's aim is to create a smooth flow for the patient through the clinical departments, ensuring proper medical care, while optimizing costs.

In developed countries the clinical departments are highly specialized, having dedicated units to specific medical areas: pediatrics, neonatology, cardiology, oncology, neurology, etc. In developing countries, on the other hand, the clinical departments tend to be less specialized and handle a wide range of conditions: general medicine, etc. This approach is strictly related to resource constraints. Due to better funding, the resources allocated for cutting-edge treatment and comprehensive patient care are higher in developed countries. In developing countries, the constraints lead to a limited service, relying on external aid and non-governmental organizations. The discrepancies go even further inside the countries. The funding and specialized personnel are clustered around big cities, leaving certain areas within that same country uncovered. Besides funding, a major difference regards health insurance and the patients' access to care. Access to care is uneven, and a significant number of patients lack health insurance. In some cases, out-of-pocket expenses are a barrier, and this fact affects how healthcare is prioritized and delivered.

Differences appear in integrating hospital services with other healthcare facilities such as primary care, outpatient services, and rehabilitation. While these departments are smoothly integrated in developed countries, in developing ones there are a lot of gaps, and the coordination of patients between the clinical departments can be quite challenging.



S. Belciug, *The Hospital Manager's Guide to Artificial Intelligence*, Intelligent Systems Reference Library 263, https://doi.org/10.1007/978-3-031-80314-7_2

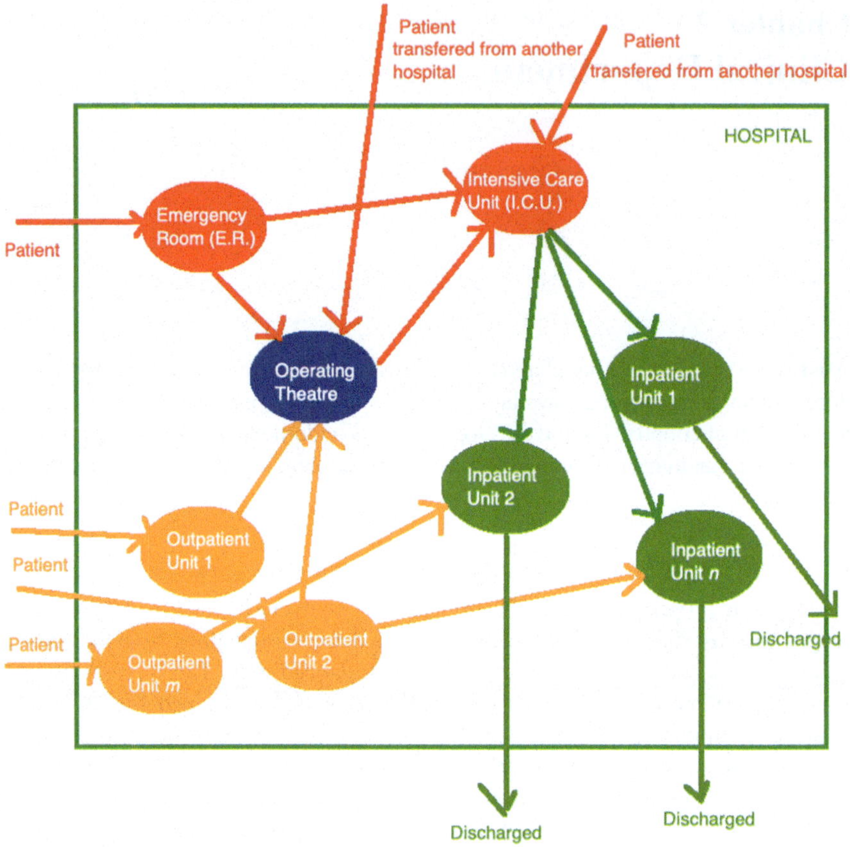


**Fig. 2.1** Patient flow inside a hospital

In Fig. 2.1 we present an example of a patient flow in a hospital. From this flowchart we can see that the manner in which the clinical departments are displayed in a healthcare facility resembles a very complex queuing system.

Queuing theory provides solutions to the queueing induced frustration. How? By applying mathematical models (the ones we have discussed in Chap. 1) to evaluate the queues and optimize their efficiency. In a hospital this means identifying the optimal number of beds, medical staff, medical supplies, number of Operating Rooms (O.R.), etc., required to provide the best care possible efficiently affecting positively the patients' waiting times. We use mathematical models to find a balance between patient care and resource constraints.

The only issue with queuing theory is that it involves the use of probability theory and mathematics. This fact might scare off the average computer scientist or hospital manager. That is why, in this book, we are going to present all these concepts in a friendly manner using a lot of examples that will show how the concepts and terminologies are interconnected. Let us see how we can make the management of clinical departments "smarter".

## 2.1 Smart Clinical Department Management

### *Queuing Theory models*

A queuing theory model studies the concept of waiting lines or queues. Its purpose is to create a system that optimizes the queue taking into account different factors and constraints. Figure 2.2 presents a diagram of a queuing theory model.

In a queuing system there are several factors that need to be defined and determined: the probability of an arriving patient needing to wait until be attended by a doctor or nurse, the average patient service time, and the average queue length (in the case where queues are permitted). Technically, a queue is formed of three elements:

- The patients that arrive according to an arrival pattern and who need medical care.
- The patients that arrive at the hospital facility and have to wait in one or multiple queues before they are seen by a doctor.
- The medical service: diagnosis, treatment, admission, etc.

### *Poisson processes*

If we have a number of random points that form a Poisson process, then the number of points in a region that has a predetermined size is a random variable that is governed by the Poisson distribution. The points occur independently of one another, [1]. Both the process and the distribution are named after Simeon Denis Poisson, a French mathematician. He discovered the distribution, after he analyzed the number of deaths by horse-kick in the Prussian army in the nineteenth century. These events were considered as being rare.

In general, there are two types of arrivals: deterministic and random. If the arriving process is deterministic then the patients follow a pattern as they arrive at the medical facility (e.g. 3 patients come into the clinic every 60 s), otherwise, the patients do

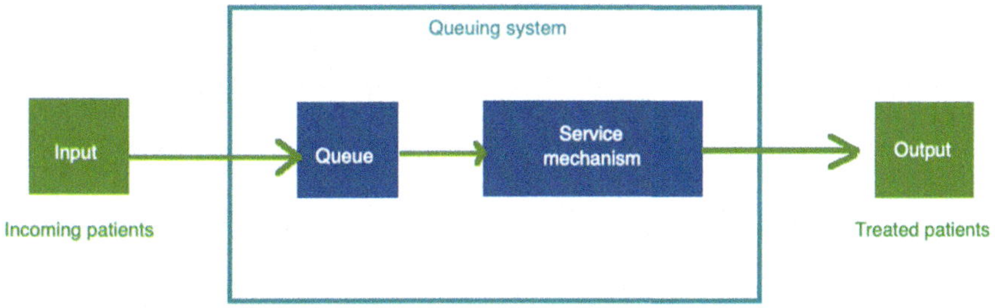


**Fig. 2.2** Queuing theory model

not follow a pattern to arrive at the hospital. Luckily enough, if three conditions are a priori fulfilled, then the Poisson distribution described in Chap. 1, can describe this random arrival process. The prerequisites are:

- The arrival process must be *continuous*, implying that at least one patient must arrive in a given timeframe.
- The arrival process must be *stationary*, implying that for a given time frame the probability of a patient arriving is equal to the probability of a patient to arrive in all the timeframes that have the same length as the given one.
- The arrival process must be *independent*, implying that the arrival of a patient is not influenced or does not influence the arrival of another patient.

Before we dive into some queuing theory examples, we need to explain with a concrete example the definition of a stochastic process.

Let us consider the following example of a queue that is formed at the general practitioner (GP) office, where a waiting line is formed for people that want their prescription renewed. In Table 2.1 we present some artificially generated observations of patients' arrivals over the course of two hours.

**Table 2.1** Observations of the moments of patients arriving at the GP's office

| Patient no. | Inter arrival time | Cumulative arrival time | Patient no | Inter arrival time | Cumulative arrival time |
|---|---|---|---|---|---|
| 1 | 1.89 | 0.03 | 21 | 1.87 | 0.89 |
| 2 | 2.18 | 0.07 | 22 | 10.92 | 0.89 |
| 3 | 2.80 | 0.11 | 23 | 3.12 | 1.07 |
| 4 | 3.31 | 0.17 | 24 | 1.88 | 1.12 |
| 5 | 1.49 | 0.19 | 25 | 0.74 | 1.15 |
| 6 | 4.06 | 0.26 | 26 | 0.90 | 1.16 |
| 7 | 2.64 | 0.31 | 27 | 2.26 | 1.18 |
| 8 | 1.41 | 0.33 | 28 | 2.73 | 1.22 |
| 9 | 4.36 | 0.40 | 29 | 0.40 | 1.26 |
| 10 | 1.41 | 0.43 | 30 | 3.28 | 1.27 |
| 11 | 2.34 | 0.46 | 31 | 0.21 | 1.32 |
| 12 | 1.98 | 0.50 | 32 | 5.50 | 1.33 |
| 13 | 0.21 | 0.50 | 33 | 1.17 | 1.44 |
| 14 | 0.35 | 0.51 | 34 | 9.12 | 1.59 |
| 15 | 0.38 | 0.51 | 35 | 2.67 | 1.63 |
| 16 | 0.90 | 0.53 | 36 | 0.17 | 1.64 |
| 17 | 15.04 | 0.78 | 37 | 12.23 | 1.84 |
| 18 | 0.53 | 0.79 | 38 | 9.32 | 2.00 |
| 19 | 2.68 | 0.83 | 39 | 5.72 | 2.09 |
| 20 | 1.32 | 0.85 | 40 | 1.59 | 2.12 |

Figure 2.3 is a graph that shows the number of patient data that falls in each time interval. Whereas Fig. 2.4 shows the cumulative number of arriving patients over the time axis. This is the sample function of a stochastic process.

Note that at any given point in time the laws of probability can describe a stochastic process, [2]. Technically, this implies that we can describe a random situation by the

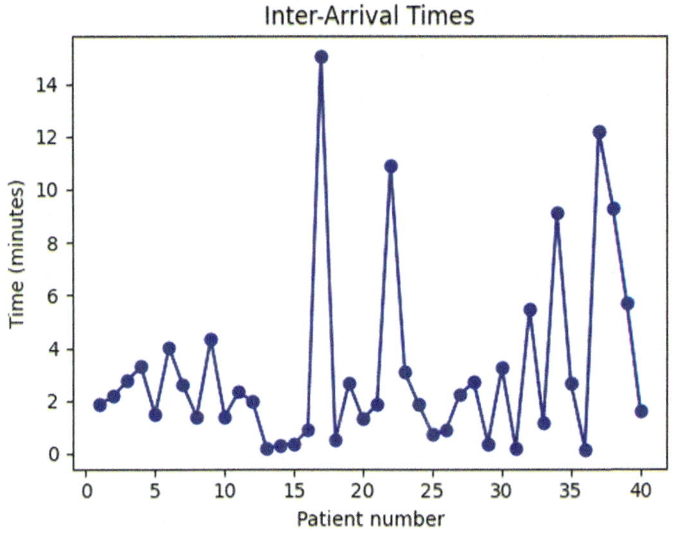


**Fig. 2.3** Interarrival times of arriving patients

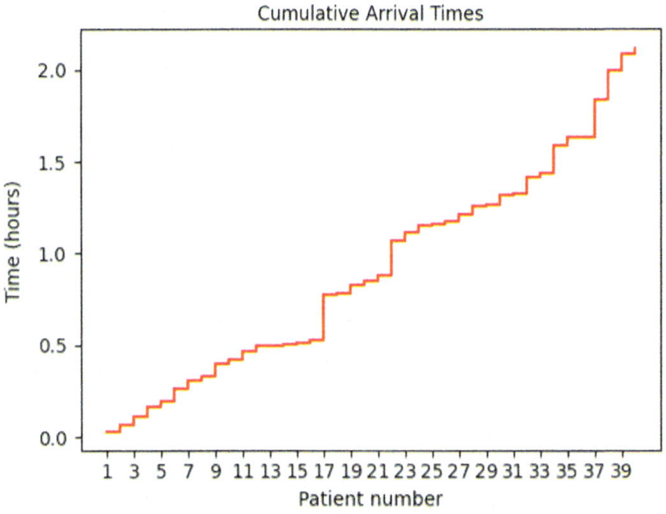


**Fig. 2.4** The cumulative number of arriving patients over the time axis

law of probability if we specific the time, $t \geq 0$. A queue that is formed at the GP office for prescription renewal is such an example, being a stochastic process.

Let us denote the number of arriving patients until time $t$ with $N(t)$, where $N(t) = 0, 1, 2, \ldots$, then we can say that the stochastic process shown in Fig. 2.4 as the sample function is the *counting process*, $\{N(t), t \geq 0\}$. One can mention other counting processes such as: the number of surgeries scheduled, the number of imagining tests programmed for an MRI (magnetic resonance imaging machine), etc. The Poisson process that will be described in this chapter is a counting process.

If we consider a time interval $[0, t]$, and we divide this interval in $n$ equal parts, then for each one of these small-time intervals we consider the Bernoulli trials having the probability of a patient to arrive equal to $p$, and the probability of a patient not to arrive equal to $q = 1 - p$. We can compute the probability of $k$ customers to arrive in the initial time interval $[0, t]$ as:

$$\binom{n}{k} p^k q^{n-k} = \frac{\left(1-\frac{1}{n}\right)\cdot\left(1-\frac{2}{n}\right)\cdot\ldots\cdot\left(1-\frac{k-1}{n}\right)}{k!}\cdot(\lambda t)^k\cdot\left(1-\frac{\lambda t}{n}\right)^{n-k}.$$

Hence, the equation for the arrival Poisson distribution is given by:

$$P\{N(t) = k\} = \frac{(\lambda t)^k}{k!} e^{-\lambda t}, k = 0, 1, 2 \ldots$$

where

- $\lambda$ is the number of expected patients to arrive per time unit.
- $t$ is the time interval.
- $e$ is the natural logarithm's base, $e = 2.7182818$.
- $k$ is the number of patients.

Hence, let us presume that at a GP's office on average we have 4 patients arriving between 8:00 and 9:00 in the morning to renew their prescription. We want to compute the probability of $k = 0, 1, 2$, and 3 patients to arrive at the GP between 8:00 and 8:30 in the morning.

We have the following input:

- $\lambda = 4$patients per hour
- $t = 0.5$ hours
- $\lambda t = 4 \cdot 0.5 = 2$ is the average number of patients arriving in half an hour.

We compute the probability of $k = 0$ patients' arrivals between 8:00 and 8:30 as:

$$P(X = 0) = \frac{e^{-2}\cdot 2^0}{0!} = e^{-2}\cdot\frac{1}{1} = e^{-2} \approx \frac{1}{2.71828^2} \approx \frac{1}{7.3891} \approx 0.1353$$

Therefore, the probability of 0 patients to arrive between 8:00 and 8:30 is 13.53%.

Following the same method, we compute that the probability of 1 patient to arrive during the first half an hour is:

$$P(X=1) = \frac{e^{-2}\cdot 2^1}{1!} = e^{-2}\cdot 2 = 2\cdot e^{-2} \approx 2\cdot\frac{1}{2.71828^2} \approx 2\cdot\frac{1}{7.3891} \approx 0.2707$$

or 27.07%.

The probability of 2 patients to arrive during the first half an hour is:

$$P(X=2) = \frac{e^{-2}\cdot 2^2}{2!} = e^{-2}\cdot\frac{4}{2} = 2\cdot e^{-2} \approx 2\cdot\frac{1}{2.71828^2} \approx 2\cdot\frac{1}{7.3891} \approx 0.2707$$

or 27.07%.

Finally, the probability of 3 patients to arrive during the first half and hour is:

$$P(X=3) = \frac{e^{-2}\cdot 2^3}{3!} = e^{-2}\cdot\frac{8}{6} = \frac{4}{3}\cdot e^{-2} \approx \frac{4}{3}\cdot\frac{1}{2.71828^2}$$
$$\approx \frac{4}{3}\cdot\frac{1}{7.3891} \approx 0.1804 \text{ or } 18.04\%.$$

or 18.04%.

The Poisson process is a discrete-time stochastic process, because the processes are observed at discrete time points. Besides these types of stochastic processes, we have the continuous-time stochastic processes (e.g. Brownian Motion, which is a mathematical model in physics that depicts the arbitrary movement of particles in a fluid). In this book, we are not going to discuss continuous-time stochastic processes, because there are no practical examples in healthcare management.

Nevertheless, besides the Poisson processes, we will discuss another important discrete-time stochastic process, Markov Chains.

***Discrete-time Markov Chains***

Markov chains were first introduced in 1913 by the Russian mathematician Andrei Andreevich Markov when he applied mathematics to poetry, [3]. After spending hours on studying the patterns of vowels and consonants in Alexander Pushkin's *Eugene Onegin*, Markov sent his theory to the Imperial Academy of Sciences in St. Petersburg. In his work, Markov computed the probabilities of a vowel to follow a consonant and vice versa. Through this computation he came up with the concept of transition matrix, which will be explained further on in this chapter. The transition matrix showed the dependency of each letter's occurrence on the previous letter. Even if it might seem unbelievable, Markov's work demonstrated that the occurrence of vowels and consonants in Pushkin's test was not random. They followed a pattern that could be statistically quantified. He demonstrated that linguistic structures do not have a random nature, and the appearance of a particular type of letter influences the next type of letter that is to occur.

Now we know how Markov's theory appeared. What we don't know is why. It all started with Andrei Markov's disagreement regarding Pavel Nekrasov's theory that stated that in order to apply the *Weak Law of Large Numbers*, the variables must be independent. For those who don't know what the Weak Law of Large Number states, we provide this simple yet understandable explanation: let us presume that we started collecting certain independent samples; as the sample size increases, the mean of the samples converges to the true mean of the population, [4].

As stated above, Markov did not agree with this statement. He thought that the independence was not a necessary condition for the mean of the samples to converge to the true mean of the population. To prove that he was right, he started to define how the average of samples that involve *dependent* random variables would converge as time passes.

He developed a theory regarding how stochastic processes evolve over time. Since the processes are modeled as a sequence of states, as time passes, the process moves from one state to another with a certain probability. All the states are connected, forming a chain. Hence the name.

Having this theory in mind, in 1957, Lejaren Hiller and Leornard Isaacson composed the *Illiac Suite*, which was a composition for string quartet, composition that included music generated using Markov chains with manually specified transition probabilities, [5]. If you care to listen to it, follow this link: https://www.youtube.com/watch?v=n0njBFLQSk8 (accessed July 9, 2024). Markov chains were used by Theo Lutz to produce a set of poems using a ZUSE Z 22 mainframe for combining vocabulary using works from Kafka's *The Castle*, [6]. Besides music and literature, Markov chains can be used to generate visual art, [7]. So, why not use them in the healthcare system?

Whoever heard of Markov Chains, has definitely heard its main feature: *memorylessness*. The fact that Markov Chains are memoryless implies that the next state of the system is independent of the previous states. It is the current state that matters. For example: patients' arrivals at healthcare facilities, the outcomes of surgeries performed on different patients, etc. Therefore, we must keep in mind, that in a Markov chain, as we move along the chain, the only state that matters for the next state is the current one. The transition between states is dependent on the state of the process before the transition occurred. Returning to Markov's theory, that the random variables do not need to be independent, by putting all these together, he proved that as long as all the states from the chain can be reached, the probability of moving to certain state will converge to a single stable value as more time passes.

Above we gave different practical examples of how to use Markov chains to produce poetry, music, and visual arts. But let's think of an example that regards hospital management. Let us presume that we are in the ER. As a patient comes to the ER she/he/they must wait for triage. Obviously, there are chairs in the waiting room, where the patient can sit while waiting. The number of chairs is fixed, but how many of these are available at any moment in time depending on different variables:

- Time of the day: the busiest times in the ER are in the late afternoon and evening hours, from 12 p.m. till 9 p.m. One possible explanation for the increased influx of patients is the fact that at those hours primary care offices or urgent care centers are closed. People with both urgent and non-urgent medical issues are seeking care at the ER. Figure 2.5 depicts a pie chart showing the ER times. To plot it we have used the data from www.figure1.com.

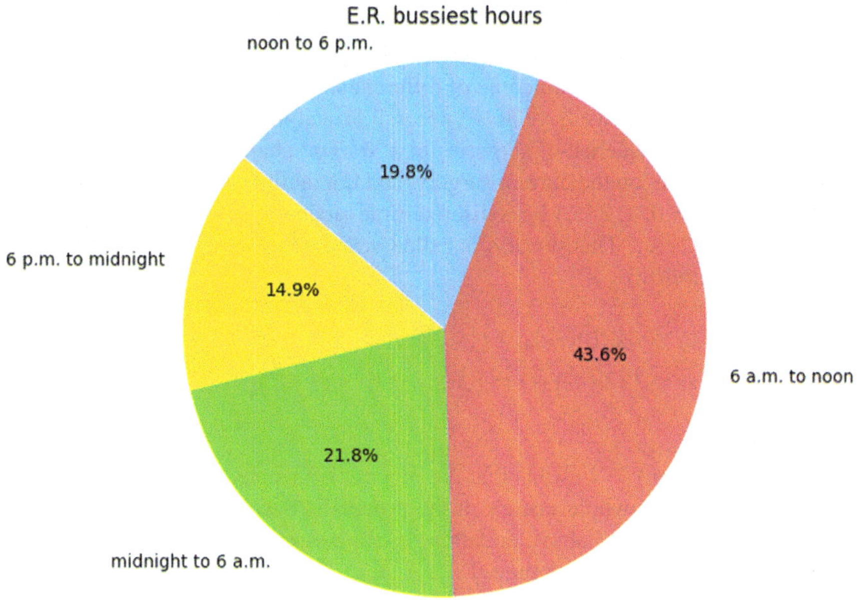


**Fig. 2.5** Busiest ER times

- Day of the week: according to www.immediatecarewestmont.com (accessed July 9, 2024), the busiest day of the week in the ER is Monday. A potential explanation is the one that most people that get sick over the weekend wait until Monday to seek care, leading to this surge of patients. After Mondays, the next busiest days are Saturday and Sunday, due to accidents and injuries that happen during recreational activities during the weekend. The middle of the week, Wednesday to Friday, are the less busy days.
- Number of available chairs in the waiting room: in general, people that do not present a true real emergency (e.g. stomachache, fever, etc.) leave the ER when they see it is too crowded.

The time of day and day of the week might be independent from each other, but the number of available chairs is dependent on the other two.

We can use Markov chains to make long-term predictions, and get answers to different questions about the future state of the process:

- In what state is the process going to be after $n$ steps of time?
- If our process goes from state $i$ to state $j$ in $n$ steps, what is the probability that it will follow a specific pattern?

In our example with the waiting room chairs, we could ask the following questions:

- What will be the rate of chair availability 2 h from now?
- How likely is that the rate of occupied chairs to be 60% now, and then 30% in the next 4 h?

As we can see so far, this type of mathematical models describes a sequence of events that might happen, events that follow the above stated Markov property. For instance, let us consider the outcome of different surgeries. Let us presume that the outcome is biased, and we have 80% successful outcome, and 20% failures. If we have a continuous sequence of surgeries in a Markov chain, each value success or failure depends only on the current surgery, and not on the previous one.

Let $\{X(n), n = 0, 1, 2 \ldots\}$ be a discrete-time stochastic process having as state space $i = 0, 1, 2, \ldots$. This process is called a *discrete-time Markov chain* with a *transition probability* $p_{ij}$ if:

$$\begin{aligned} P\{X(n+1) = j | X(0) = i_0, X(1) = i_1, \ldots, X(n-1) = i_{n-1}, X(n) = i\} \\ = P\{X(n+1) = j | X(n) = i\} = p_{ij} \end{aligned}$$

for all $i_0, i_1, \ldots i_{n-1}, i, j, n$.

The notation $X(n) = i$ means that the process is in state $i$, for $i = 0, 1, 2 \ldots$ at a certain time $n$. The whole history of the process is described by the conditional probability written in the above equation. It can be easily seen that the process is independent by the history at time $0, 1, \ldots, n-1$, and is dependent only on the current state. As the process changes from one state to the next a *state transition* takes place. The probability that is associated with each transition is the *transition probability* $p_{ij}$, [8].

Imagine a Markov chain with $N$ states. We denote the matrix $\mathbf{P} = [p_{ij}]$ as the *transition probability matrix*, having:

$$p_{ij} \geq 0, \quad \sum_{j=0}^{\infty} p_{ij} = 1$$

for $i, j = 0, 1, 2, \ldots$. Each cell from this matrix represents the probability of the system to transition from the state $i$ to the state $j$. The transition matrix is a stochastic matrix, meaning that by adding the values of each row the obtained sum is 1. Why? Because each row represents its individual probability distribution.

***Chapman-Kolmogorov Equation***

We can describe the probabilistic comportment of a Markov chain $\{X(n), n = 0, 1, 2 \ldots\}$ using the initial probability of the states and the transition probabilities, [2]:

$$P\{X(0) = i_0, X(1) = i_1, \ldots, X(n) = i_n\}$$

$$P\{X(n) = i_n | X(0) = i_0, \ldots, X(1) = i_1, \ldots, X(n-1) = i_{n-1}\}$$

$$P\{X(0) = i_0, X(1) = i_1, \ldots, X(n-1) = i_{n-1}\}$$

$$= p_{i_{n-1},i_n} \cdot P\{X(0) = i_0, X(1) = i_1, \ldots, X(n-1) = i_{n-1}\}$$
$$= \ldots$$
$$= p_{i_{n-1}i_n} \cdot p_{i_{n-2}i_{n-1}} \cdot \ldots \cdot P\{X(0) = i_0\}$$

in which $P\{X(0) = i_0\}$ is the initial probability.

We denote with $\pi(0)$ the initial distribution, having:

$$\pi(0) = [\pi_0(0), \pi_1(0), \ldots]$$

with

$$\pi_j(0) = P(X(0) = j\} \geq 0, j = 0, 1, 2 \ldots$$

having the sum:

$$\sum_{j=0}^{\infty} \pi_j(0) = 1.$$

By using the initial distribution $\pi(0)$ and the transition probability matrix, we can compute all the transition probabilities. Mathematically speaking, we denote:

$$p_{ij}^n = P\{\ X(n+m) = j|X(m) = i\}$$

the $n$th transition probability in which we will have the process state $j$ at time $n + m$, knowing that the current state is $i$, at time $m$. This probability is dependent on the time duration $n$, and independent of the present time $m$. We can compute $p_{ij}^n$ using $p_{ij}$, where we assume that:

$$p_{ij}^0 = 0 \quad \text{for} \quad i \neq j.$$

Therefore, we can compute the $n^{\text{th}}$ transition probability by summing up all the intermediate states $k$ at time $l$, while moving to the state $j$ from $k$, at the remaining time $n - r$:

$$p_{ij}^n = \sum_{k=0}^{\infty} p_{ik}^r \cdot p_{kj}^{n-r}.$$

The above equation is known under the name *Chapman-Kolmogorov equation.* We can rewrite this equation under a matrix form, for easier computations:

$$\mathrm{P}^{(n)} = \left[p_{ij}^n\right]$$

$$P^{(n)} = P^{(r)} \cdot P^{(n-r)}.$$

From a computational point of view, knowing that $P^{(1)} = P = [p_{ij}]$, we can recursively compute $P^{(n)}$ as:

$$P^{(n)} = P^{(1)} \cdot P^{(n-1)} = P \cdot P^{(n-1)} = P^2 \cdot P^{(n-2)}$$
$$= \cdots = P^n$$

To better understand the theory, let us take the following example of how to compute the Chapman-Kolmogorov equation.

The above sequence of matrix multiplications using previously computed transition matrix is known under the name of *recursion*. Each resulting matrix is named the *power of the transition matrix.*

Let's presume that we are managing a small healthcare facility that deals with infectious diseases. We want to see if we can determine what will happen tomorrow, if we know what happened today. We have three possibilities:

- If today all the patients that presented themselves at the clinic were diagnosed with type A flu, there is a 30% chance that tomorrow all the patients will be diagnosed with type A flu, 40% chance that they will be diagnosed with type B flu, and 30% chance that they will be diagnosed with the common cold. You can see that if we add the probabilities, the sum is 100%.
- If today all the patients will be diagnosed with the common cold, there is a 20% chance that tomorrow all the patients will be diagnosed with the common cold, 50% chance that they will be diagnosed with type A flu, and 30% that they will be diagnosed with type B flu.
- If today all the patients were diagnosed with type B flu, 50% change tomorrow they will be diagnosed with type A flu, 20% with the common cold, and 30% with type B flu.

The above information is known as state transition, and we need to convert it into a state transition matrix. Table 2.2 shows how the state transition matrix would look like in our case.

Let us see how we can compute what is the probability that all the patients that will present themselves to the clinic tomorrow to be diagnosed with a certain disease. We have the following possible states: type A flu, common cold, and type B flu. If we were to say that today all the patients were diagnosed with common cold, then the current state is [010], and our transition matrix is:

**Table 2.2** State transition matrix example for diagnosing unit

| | Type A flu | Common cold | Type B flu |
|---|---|---|---|
| Type A flu | 0.3 | 0.3 | 0.4 |
| Type B flu | 0.5 | 0.2 | 0.3 |
| Common cold | 0.5 | 0.2 | 0.3 |

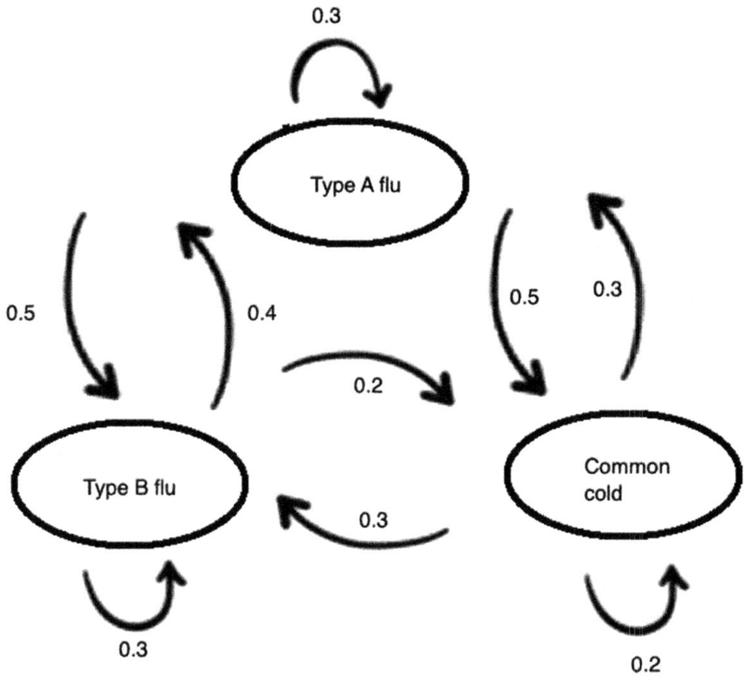


**Fig. 2.6** The graph that corresponds to the state transition matrix

$$\begin{bmatrix} 0.3 & 0.3 & 0.4 \\ 0.5 & 0.2 & 0.3 \\ 0.5 & 0.2 & 0.3 \end{bmatrix}.$$

The transition matrix can be transformed into a graph (Fig. 2.6).

By performing a matrix multiplication, we obtain $[\,0.5\ 0.2\ 0.3\,]$, which can be translated as: the probability from common cold to type A flu is 0.5, common cold to common cold is 0.2, and common cold to type B flu is 0.3.

But what if we want to compute the probability for the next 5 days? We just need to repeat the step presented above by multiplying the new state matrix [0.50.20.3] with the transition matrix:

$$[\,0.5\ 0.2\ 0.3\,] \times \begin{bmatrix} 0.3 & 0.3 & 0.4 \\ 0.5 & 0.2 & 0.3 \\ 0.5 & 0.2 & 0.3 \end{bmatrix} = [\,0.33\ 0.38\ 0.38\,]$$

The resulted matrix $[\,0.33\ 0.38\ 0.38\,]$ represents the probability for the third day. The fourth day probability is obtained by multiplying the new resulted matrix with the state transition matrix. We get:

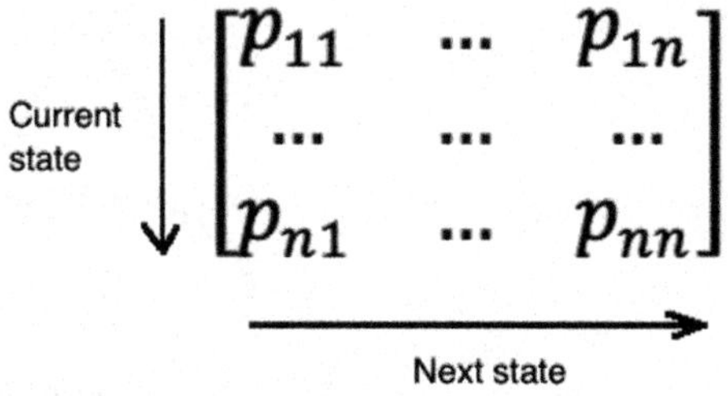


**Fig. 2.7** Transition matrix for a Markov chain with $n$-states

$$[\,0.33\ 0.38\ 0.38\,] \times \begin{bmatrix} 0.3\ 0.3\ 0.4 \\ 0.5\ 0.2\ 0.3 \\ 0.5\ 0.2\ 0.3 \end{bmatrix} = [\,0.479\ 0.251\ 0.36\,]$$

Finally, for the fifth day, we compute the new resulted matrix with the transition matrix and obtain:

$$[\,0.479\ 0.251\ 0.36\,] \times \begin{bmatrix} 0.3\ 0.3\ 0.4 \\ 0.5\ 0.2\ 0.3 \\ 0.5\ 0.2\ 0.3 \end{bmatrix} = [\,0.4492\ 0.2659\ 0.3749\,]$$

And so on. We can continue using the same algorithm for a number of $n$ days.

If we would have a Markov chain with $n$ states, then the transition matrix for this case would look like the one depicted in Fig. 2.7.

What if we would think of the management of clinical departments as we think of playing chess. Plan ahead. Imagine what will happen next. Ask ourselves how will a patient move around our hospital? Which department will the patient be admitted next? Etc. One potential solution would be to think of the clinical departments as states in a Markov Chain. The next department in which the patient will be admitted depends only on the current department in which the patient is admitted. So, we think of the hospital as a finite-state, discrete-time, homogeneous Markov chain. In what follows we will explain our idea in two separate ways: using discrete graphs and using matrix analysis. The characteristics of the finite-state Markov chains allow us to switch between the numerical and the graphical representations.

At first, we shall think of the a patient' movement through a hospital, from one clinical department to the next, as a *random walk*. Technically, a random walk is just a series of *discrete steps* that an object, in our case a patient, takes in a direction, in our case in which department will the patient be admitted next. We will determine the next department and patient admission in each step probabilistic. In probability theory the random walk is a *random process*.

Let us consider the following example: we have a hospital with 7 clinical departments. We are going to simulate the stochastic process of how a patient moves around the hospital by using the next transition matrix:

$$\begin{bmatrix} 0.2 & 0.3 & 0.0 & 0.0 & 0.0 & 0.4 & 0.1 \\ 0.0 & 0.0 & 0.0 & 0.0 & 0.2 & 0.2 & 0.4 \\ 0.0 & 0.0 & 0.3 & 0.2 & 0.5 & 0.0 & 0.0 \\ 0.0 & 0.2 & 0.0 & 0.2 & 0.4 & 0.2 & 0.0 \\ 0.0 & 0.0 & 0.0 & 0.5 & 0.4 & 0.0 & 0.1 \\ 0.0 & 0.0 & 0.0 & 0.3 & 0.0 & 0.3 & 0.5 \\ 0.0 & 0.3 & 0.0 & 0.3 & 0.0 & 0.0 & 0.4 \end{bmatrix}$$

If our patient starts in the clinical department number 4, using the above transition matrix and theory, the patient will transition throughout the clinical departments using the following random walk steps $[4, 2, 6, 6, 7, 2, 6, 6, 6, 4, 5]$. The graphical representation of this simulated random walk is presented in Fig. 2.8, where you can see how the patient moves around the hospital. We have highlighted each step with numbers and if the patient is kept in the same clinical department after a procedure, we have marked this step as an arc that goes out one node and enters the same node. Having this example in mind, one can adjust the Markov chain scenario by modifying the state transition matrix and observing different behaviors and patterns.

Let us return a little bit to what Markov wanted to prove. He wanted to show that if we have a large enough number of iterations, the transition probabilities will indeed converge to a certain value and will remain unchanged. Practically, this means that if we have enough iterations, the probability of ending up in any state of the chain is the same, no matter in which state the process begins. We say that when the chain reached this point, then the transition probabilities reached a *steady state*.

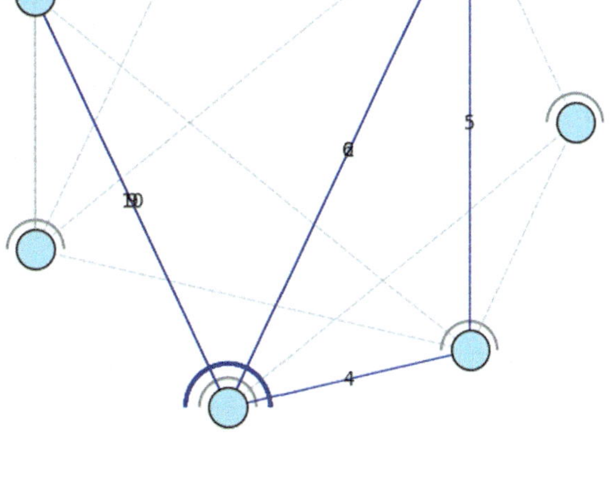


**Fig. 2.8** Random Walk graphical representation of a patient moving from one clinical department to another

We are going to simulate the way a patient is moved through different clinical departments, to see whether our Markov chain will reach a steady state. The states represent the following clinical departments:

- State 0: ER.
- State 1: Surgery department.
- State 2: ICU.
- State 3: Internal Medicine.
- State 4: Radiology.
- State 5: Laboratory.
- State 6: Oncology.
- State 7. Radiotherapy.

The simulated transition matrix is the following:

$$\begin{bmatrix} 0.3\ 0.2\ 0.1\ 0.1\ 0.1\ 0.1\ 0.1 \\ 0.1\ 0.3\ 0.2\ 0.1\ 0.1\ 0.1\ 0.1 \\ 0.1\ 0.1\ 0.3\ 0.2\ 0.1\ 0.1\ 0.1 \\ 0.1\ 0.1\ 0.1\ 0.3\ 0.2\ 0.1\ 0.1 \\ 0.1\ 0.1\ 0.1\ 0.1\ 0.3\ 0.2\ 0.1 \\ 0.1\ 0.1\ 0.1\ 0.1\ 0.1\ 0.3\ 0.2 \\ 0.1\ 0.1\ 0.1\ 0.1\ 0.1\ 0.2\ 0.3 \end{bmatrix}$$

We are going to presume that the patient came to the hospital and went to the ER. Hence the initial state of the process is: [ 1.0 0.0 0.0 0.0 0.0 0.0 0.0 ]. By applying Chapman-Kolmogorov equations we get:

$$\text{Current state}:\ [\,1.\ 0.\ 0.\ 0.\ 0.\ 0.\ 0.\,]$$

*Step 1:*

$$[\,1.\ 0.\ 0.\ 0.\ 0.\ 0.\ 0.\,]$$

*Step 2:*

$$[\,0.3\ 0.2\ 0.1\ 0.1\ 0.1\ 0.1\ 0.1\,]$$

*Step 3:*

$$[\,0.16\ 0.17\ 0.14\ 0.13\ 0.13\ 0.14\ 0.13\,]$$

*Step 4:*

$$[0.132\ 0.15\ 0.145\ 0.14\ 0.139\ 0.154\ 0.14]$$

*Step 5:*

$$[0.1264\ 0.1432\ 0.144\ 0.1425\ 0.1418\ 0.1587\ 0.1434]$$

*Step 6:*

$$\left[0.12528\ 0.14128\ 0.14312\ 0.1429\ 0.14261\ 0.16026\ 0.14455\right]$$

*Step 7:*

$$[0.125056\ 0.140784\ 0.142752\ 0.142892\ 0.142812\ 0.160768\ 0.144936]$$

*Step 8:*

$$[0.1250112\ 0.1406624\ 0.1426288\ 0.1428536\ 0.1428516\ 0.1609284\ 0.145064]$$

*Step 9:*

$$[0.12500224\ 0.1406336\ 0.142592\ 0.1428336\ 0.14285568\ 0.16097724\ 0.14510564]$$

*Step 10:*

$$[0.12500045\ 0.14062694\ 0.14258176\ 0.14282592\ 0.1428545\ 0.16099158\ 0.14511885]$$

*Step 11:*

$$[0.12500009\ 0.14062543\ 0.14257905\ 0.14282336\ 0.14285349\ 0.16099565\ 0.14512293]$$

*Step 12:*

$$[0.12500002\ 0.1406251\ 0.14257835\ 0.14282258\ 0.14285303\ 0.16099677\ 0.14512415]$$

*Step 13:*

$$[0.12500002\ 0.1406251\ 0.14257835\ 0.14282258\ 0.14285303\ 0.16099677\ 0.14512415].$$

You might have asked yourself if all Markov chains reach a steady-state. The answer is no. Only regular Markov chains have this property. If a Markov chain does not converge, it contains a periodic pattern in it. In general, we define a regular Markov chain if a power of its transition Matrix contains only non-zero, positive values. If it isn't regular, instead of reaching a steady-state, the chain will produce the same transition probabilities from time to time.

There is a way to test if your Markov chain is regular. If you have a $n \times n$ transition matrix, then you can check the powers of the transition matrixes. If they have positive, non-zero values, then your Markov chain is regular.

In our example with the patient transitioning between 7 departments, let us compute the likelihood of following a specific path: ER, Surgery department, ICU, Internal Medicine, Radiology, Laboratory, Oncology, Radiotherapy?

To find the answer we must look at the Markov chain. Recall that each single step in the chain corresponds to a conditional probability. Therefore, the likelihood to follow a specific path is the sum of all conditional probabilities of that path. In our case the answer is: 0.000012911113576316172.

If we want to compute the likelihood of a patient to reach any department in step 4, we need to take note that to answer the last question we knew an important detail: in which department the patient arrived first. If we presume that the patient can enter the hospital from any department, and not just the ER, then we would say that the starting state of our Markov chain follows a discrete uniform distribution, and we should compute the probabilities of being admitted to any department considering this information. The likelihood of reaching any department in step 4 is computed by multiplying the starting probability vector by the transition matrix at power 4.

So far, so good. We have started this chapter by learning a little bit of the differences between clinical departments in developed and developing countries, and we have reached the conclusion that the way a patient moves through the hospital resembles a lot to a queuing system. We did not dive any further into different queueing models, but we did start learning about stochastic process, Poisson processes, and Markov chains. In the following chapters we shall discuss how queuing models and AI models can be used to make hospital management smarter.

## References

1. Chiu, S.N., Stoyan, D., Kendal, W.S., Mecke, J.: Stochastic Geometry and its Applications. Wiley (2013)
2. Osaki, S.: Applied Stochastic System Modeling. Springer-Verlag (1992)
3. Basharin, G.P., Langville, A.N., Naumov, V.A.: The life and work of A.A. Markov. Linear Algeb. Appl. **386**, 3–26 (2004)
4. Eddy, S.: What is a hidden Markov model? Nat. Biotechnol. **22**, 1315–1316 (2004)
5. Lejaren, H., Isaacson, L.M.: Experiment Music; Composition with an Electronic Computer. McGraw-Hill (1957)
6. Lutz, T.: Stochastic Texts (1959). https://stuttgarter-schule.de/lutz_schule_en.htm
7. Gristwood, S.: Hiroshi Kawano (1925–2012): Japan's pioneer of computer arts. Leonardo **52**(1), 75–80 (2019)
8. Kemeny, J.G., Snell, J.L.: Finite Markov Chains. Springer-Verlag (1976)

# Chapter 3
# Laboratory

**Abstract** The lab is where patients tests are performed and analyzed. The faster the tests are performed, the faster appropriate treatment is given to the patient, hence we need to optimize how the laboratory works. In this chapter we shall present some AI models that are able to do so, along with practical examples.

Whenever someone goes to a healthcare facility, a doctor performs a physical check and talks to the patient regarding its medical history and other factors that might influence her/his/their health. Even if the patient is telling the truth about her/his/their symptoms, it is difficult to quantify the severity of the disease, or even its existence, since everybody has a different pain threshold. Another misleading factor is represented by "Doctor Google" or "Doctor ChatGPT". Patients are surfing the web asking questions regarding their symptoms and start diagnosing themselves. Most people after reading other related symptoms might start imagining things. People often lie about their medical or personal history, about their job or housing, etc., because they sometimes feel ashamed that they did drugs or are unemployed or live in a very dirty house. Thus, the only things that a doctor can trust are the symptoms that can be seen (i.e. increased heart rate, rash, stroke, etc.) or tests (blood tests, imaging, etc.).

The workflow of performing a non-imagistic lab test is depicted in Fig. 3.1.

The diagnostic process begins with the doctor ordering different tests based on the clinical symptoms presented by the patient. At this stage we can encounter the following situation: over-testing.

Over-testing: why more tests could cause more harm than good? Well, the first noticeable disadvantage is the fact that is time-consuming. All the steps presented in Fig. 3.1 take time. If a test is not necessary, then that time is lost for that lab doctor, for another patient, for the current patient, and for the doctor that orders them. Another disadvantage is the billing. Tests cost money, whether that money comes from an insurance company or out of pocket. That money could be spent elsewhere. The third reason that we are going to mention is the patient's mental status. When a person sees the amount of tests ordered by a doctor, the mental status is altered making the patient stressed, heart rate increased, rapid breathing, etc. None of the above are

 
S. Belciug, *The Hospital Manager's Guide to Artificial Intelligence*, Intelligent Systems Reference Library 263, https://doi.org/10.1007/978-3-031-80314-7_3

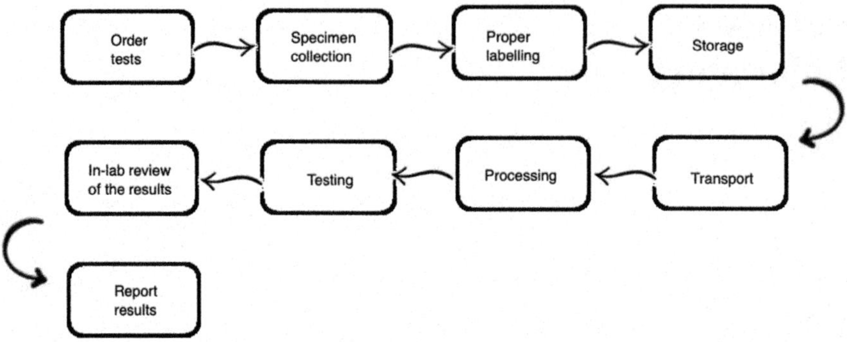


**Fig. 3.1** Performing lab tests workflow

in the benefit of the patient. And last, but not least, in some situations diagnosing diseases might lead to worst outcomes than not diagnosing them. Not doing any tests or interventions when you guess there is a disease present is known under the name of *watchful waiting* or *watch and wait*. Watchful waiting is recommended in the situation where there is a high likelihood that the issue will resolve on its own (ear infection in children, uterine polyps, abdominal pains, depression, prostate cancer in older patients etc.).

Having all this in mind, we cannot help to wonder why so many clinicians over-test? There are at least five reasons why this phenomenon happens:

1. Practicing defensive medicine: doctors are afraid that something might go wrong with their patient, and they prefer to be on the safe side. This is why so many unnecessary antibiotics are prescribed every day, leading to a future global problem: antibiotic resistance bacteria.
2. Believing that by ordering more tests they will detect subclinical diseases. A clinical disease has signs and symptoms, which a doctor can recognize. A subclinical disease is a silent disease, that attacks the body before the onset of symptoms (e.g. diabetes, hypothyroidism, rheumatoid arthritis, genetic diseases, cancer).
3. Lacking knowledge or confidence: even if this idea might seem outrageous, it is correct and it exists. Some doctors are not fit to be doctors, and they hide their insecurities, lack of experience and know-how by over-testing, [1].
4. Patient's expectations: every patient that feels sick wants answers and treatment immediately. This is most common in parents of sick children. Letting a disease run its course is very hard, and requires a lot of nerve, courage, patience, and trust in your doctor. If a doctor uses as approach watchful waiting, then the patient might see that as a mistake. In this case, the patient will leave that clinic and go to another doctor that orders tests and prescribes drugs.
5. Profit: as awful as this might sound is as real as possible. At the end of the day, clinics and hospitals are businesses, and businesses need to survive. How? By bringing more patients, performing more procedures and interventions, and you have guessed correctly, more tests. If patients are insured, doctors believe that

it's a case of "no harm, no foul": the money is not out of pocket, it comes from the insurance company. This does not mean that profit based on patient's out of pocket money is not possible, [2].

All these reasons can and will be eliminated using AI How? We will see in the next section. Until then, we still have different ideas to discuss. For instance, false positives. When ordering unneeded test, few clinicians consider the low gain/high cost of the patient, and the considerable amounts of false positives. Besides the amount of money spent, abnormal test results that later on are proved to be incorrect, produce a huge amount of stress and anxiety for the patient, not to mention further unnecessary testing, [3, 4].

In a survey, 91% of clinicians stated that they order more tests than necessary to protect themselves against malpractice suits, [5]. Unfortunately, ordering tests does not lead to avoiding mistakes and lawsuits. As stated above, false positives can lead to unnecessary procedures that might go totally wrong, [6, 7].

It is obvious that over-testing is rapidly becoming a growing concern in the healthcare system, [8–10]. Several movements have started addressing this issue: Choosing Wisely, Less is More, and Too Much Medicine, [11–14]. We propose another approach: using AI to prevent over testing. Each time a doctor orders tests that appear not to be crucial, a red flag should be raised, and the hospital manager should take appropriate decision: approving or disapproving the test.

## 3.1 Smart Laboratory: AI Steps In

Laboratory testing represents a crucial part of diagnosing a patient. Around 60–70% of the clinical decision is taken considering the tests results, [15, 16]. Before discussing different AI methods that might come in handy in resolving the over-testing problem, we need to explain a concept that appeared earlier in this chapter: *false positive* (FP). When it comes to medical diagnosis things are far from being perfect. Everyday, doctors and patients are faced with *false positives* or *false negatives,* (FN).

***False positive and false negative***

A *false positive* means that a patient has received a positive test result, when in fact the correct result was a negative one. For instance, a person that has a COVID-19 PCR test and it comes back positive, when in fact it should have been negative. In statistics, this type of error is known under the name of *Type I error*, or $\alpha$. If we encounter a *Type I error*, then it means that the *null hypothesis* was wrongfully rejected.

The *null hypothesis* is denoted by $H_0$, and it is the usually accepted hypothesis. Besides the null hypothesis, we have its opposite, $H_1$ the *alternative hypothesis.* In statistics, when we make a presumption, or a hypothesis, we have to test it to see whether it is correct or not. The null hypothesis is what statisticians use when they want to test their new discoveries. Theoretically, we should test it on the whole population, but practically, that is impossible. Hence, we need to choose a random sample of our population and examine that sample. To trust our results, we must make sure that the random sample matches the exact characteristics of the entire population. For instance, if the general population has 60% women and 40% men, then our sample must keep the same proportions. The same thing goes for age groups, education, etc. After we create the sample population, we formulate a hypothesis. For instance: "the lab from location A is accessed by more patients that the lab from location B". We have:

- $H_0$: states that there is no significant difference between the number of patients that go to the lab from location A and the number of patients that go to the lab from location B. The difference that was noticed was just produced by hazard.
- $H_1$: states that there are significant differences between the number of patients that go to lab A and the number of patients that go to lab B. The reported differences were not caused by hazardous situation.

The opposite of *false positive* is the *false negative*, meaning that a patient received a negative test result, when it was a positive one. A negative COVID-19 PCR test that marked the patient as being COVID-free, when in fact the patient had the disease.

Another concept is the *false positive paradox*, which appears when a test has a higher chance of being a false positive, than the probability of that disease to occur. For example: a patient has received a positive test result for a blood analysis, test that has a 99% accuracy in detecting the disease. What are the chances of that patient to actually have that disease? It depends on how common that disease really is. If it is a common disease, then the chances of having that disease increase, and get close to 99%. But if we are talking about a rare condition, then the odds drop a lot.

Let us say that a person got tested for a rare autoimmune disease, that affects 1 in 10,000 people. We know that if a patient has this autoimmune disease, then test will be 99% positive, and if the person is healthy, then the test will be 0.1% positive. What is the probability of a person to be ill, if the test is positive? Let us see what over-testing does. If the patient receives the positive test result, the doctor will start them on steroids which will destroy their immune system, which ultimately will make the patient prone to infections and other diseases, which jeopardize her/his/their life. We are going to use *Bayes' theorem* to answer our question [17].

*Bayes' theorem*: for any event $A \subseteq \Omega$, where $\Omega$ is the measure space, the probability $P(A)$ satisfies the following axioms:

*Kolmogorov's axioms*

1. $0 \leq P(A) \leq 1$
2. $P(A) = 1$
3. For any sequence of mutually exclusive events $(A_n)_n$, we have:

$$P\left(\bigcup_{n=1}^{\infty} A_n\right) = \sum_{n=1}^{\infty} P(A_n).$$

Recall, that in the second chapter we have discussed conditional probabilities. As a short recap, a conditional probability is the probability of an event to happened knowing that another event happened before. The formula for conditional probability is:

$$P(A|B) = \frac{P(A \cap B)}{P(B)}.$$

Then, using the *total probability law:*

$$P(B) = \sum_{i=1}^{n} P(B|A_i) \cdot P(A_i),$$

Bayes' formula is:

$$P(A_i|B) = \frac{P(B|A_i) \cdot P(A_i)}{\sum_{i=1}^{n} P(B|A_i) \cdot P(A_i)},$$

where $P(B) > 0, P(A_i) > 0$, for $i = 1, 2, \ldots, n$.

We can translate Bayes' formula in words using the following notations:

- $P(A_i|B)$ is the *posterior probability.*
- $P(A_i)$ is the *prior probability.*
- $P(B|A_i)$ is the *likelihood.*
- $P(B)$ is the *evidence.*

Hence, Bayes' theorem is:

$$posterior\_probability = \frac{likelihood \times prior\_probability}{evidence}.$$

Now, let's apply this theory and compute our answer: let $B$ be the event that the patient suffers indeed from a rare autoimmune disease, having $\overline{B}$ the event that the patient does not have that disease. $T$ represents the positive test. By applying Bayes' formula, we will have:

$$P(B) = 0.0001, P\left(\overline{B}\right) = 0.9999$$

$$P(T|B) = 0.99, P\left(T|\overline{B}\right) = 0.001$$

then:

$$P(B|T) = \frac{P(T|B) \times P(B)}{P(T|B) \times P(B) + P(T|\overline{B}) \times P(B)} = 0.09.$$

Here is an example of over-testing: the chance that our patient that had a positive test result for a rare autoimmune disease to have in fact the disease is 9%. Just recall what will happen if we would have started the steroid treatment for the patient.

After demonstrating how Bayes' theorem can help prevent the effects of over-testing, let us return to learn more about false positive, false negative, and the other two related statistical concepts: *true negative* (TN), and *true positive*(TP). These four parameters are used to compute the accuracy of a test, of a procedure, of a treatment, or of an AI model. A *true positive* test result means that the test result is indeed correct, the patient has that disease. On the other hand, a *true negative* result means that the test is correct, and the patient is disease free. Using the four above mentioned parameters, we compute the accuracy as:

$$accuracy = \frac{TP + TN}{TP + TN + FP + FN}.$$

The values can be tabulated in the form of confusion matrix (Figs. 3.2 and 3.3). Depending on the number of decision classes we can have a matrix of $2 \times 2$, or $m \times m$, where $m$ is the number of possible decisions. If we add the numbers that appear on the first diagonal, we obtain the correctly determined cases.

We have seen that by plotting a heatmap together with the numbers, the confusion matrix is easier to be understood.

When we are performing a test, besides the accuracy, we are interested in finding out how sensitive or specific the test is. For this, we need to compute the *sensitivity* and *specificity*. For even more information, it is recommended to compute the *negative predictive value* (NPV), and *positive predictive value* (PPV) along with the sensitivity and specificity.

Sensitivity is computed as the proportion of true positives that are identified as such, while specificity is computed as the proportion of true negatives that are identified as such.

$$sensitivity = \frac{TP}{TP + FN}$$

$$specificity = \frac{TN}{TN + FP}$$

These two measures tell us how good the test is in finding patients that have the disease and finding the patients that do not have the disease. Besides this information, we should be interested in finding out what is the proportion of patients that were tested positive and have indeed the disease, and the proportion of patients that were tested negative and are indeed disease free. For this, we use the PPV and NPV parameters, computed as:

$$PPV = \frac{TP}{TP + FP}$$

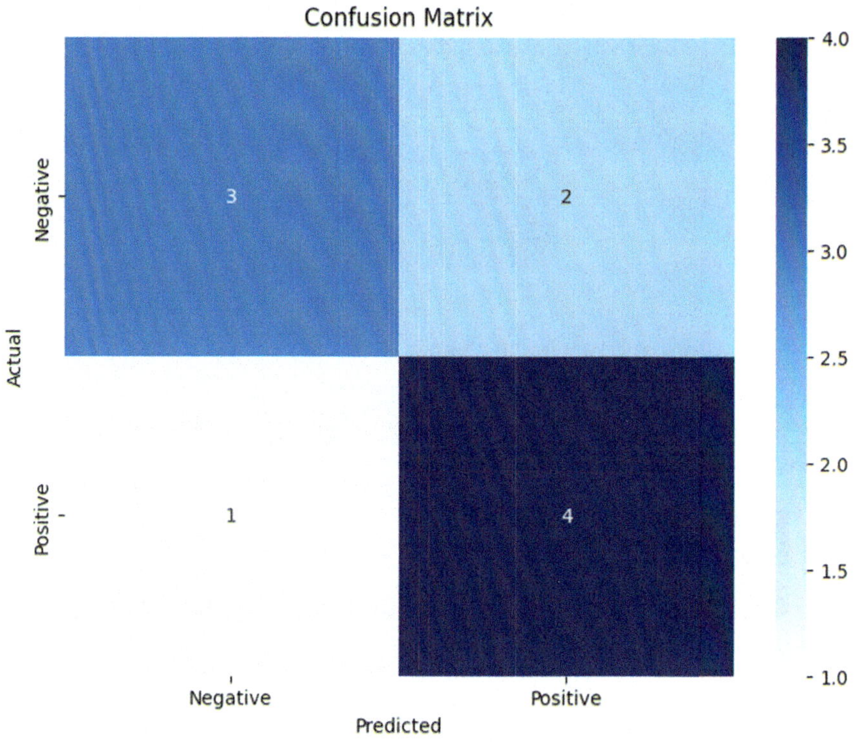


**Fig. 3.2** Confusion matrix for a two-class decision problem

$$NPV = \frac{NPV}{TN + FN}$$

For even a more thorough analysis, four other parameters can be computed:

- *False positive rate*:

$$FP_{rate} = \frac{FP}{FP + TN} = 1 - specificity$$

- *False negative rate*:

$$FN_{rate} = \frac{FN}{TP + FN} = 1 - sensitvity$$

- *Likelihood ratio positive*

$$LR+ = \frac{sensitivity}{1 - specificity}$$

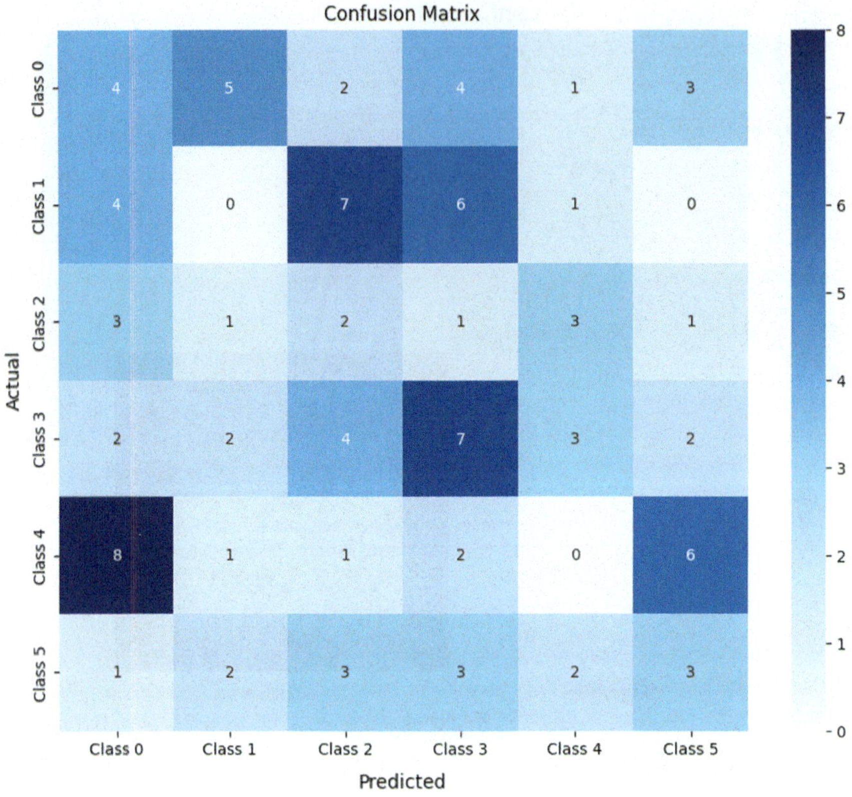


**Fig. 3.3** Confusion matrix for a six-class decision problem

- *Likelihood ratio negative*

$$LR- = \frac{1 - sensitivity}{specificity}$$

Likelihood rations (LR) combine the test's sensitivity and specificity to express a chance in odds. The positive LR computes the change in the odds of having a disease in patients that received a positive test result. For instance, a $LR+ = 10$ might indicate a tenfold increase in the odds of having a certain disease in a patient that has a positive test. If LR +'s value increases, the more informative the test gets. If LR +'s value is 1, then the odds of having the disease did not change after receiving the test results. The test is pretty much useless.

On the other hand, a negative LR computes the chance in the odds of a patient with a negative test result in having the disease. A $LR- = 0.1$ indicates a tenfold decrease in the odds of having a disease after a negative test. If the value of LR- is smaller, then the test is more informative. Again, if the value is 1, the test is useless.

Note that if we have a tenfold increase in odds that a patient has an autoimmune disease, it does not answer the patient's question: "What are the chances that I have an autoimmune disease?". Please remember that the value 10 does not mean a tenfold increase in the probability, but rather in the odds. LRs are used to project the change in the probability of having a disease before we test for that disease (pre-test probability), and the probability of having the disease after we performed the test (post-test probability). Figure 3.4 depicts a *nomogram*, which can be used to visualize LR using the pretest probability and posttest probability.

Another way we can assess the prediction results is by using the *Receiver Operating Characteristic Curves* (ROC). ROC curves were developed in World War II, after Pearl Harbor. As the Japanese planes appeared on the radar flying in the form of a flock of birds formation, the radar picked up this behavior, but was interpreted as "Don't worry about it. It is just a flock of birds". After this event, electrical and radar engineers developed the ROC curves, which have been used in healthcare research, [18–23]. Figures 3.5 and 3.6 present two plots of the ROC curve, one discrete example and one continuous. We can see that a ROC curve is actually a 2-dimensional graph that has on the X-axis the FP rate, and on the Y-axis the TP rate. By plotting the ROC curve, we can find the trade-off between costs and benefits easier.

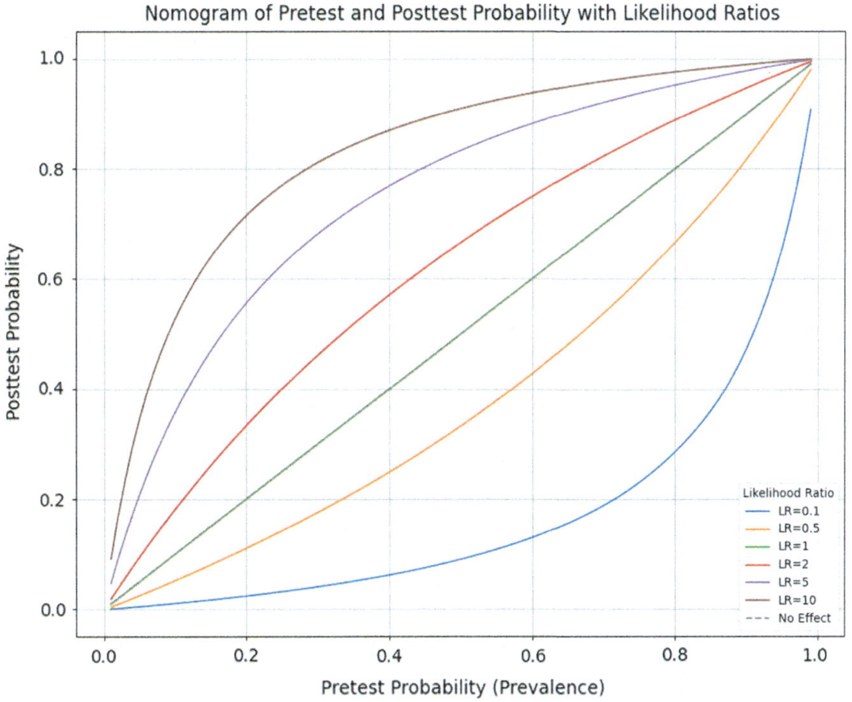


**Fig. 3.4** Nomogram illustrating Likelihood ratio, pretest and posttest probabilities

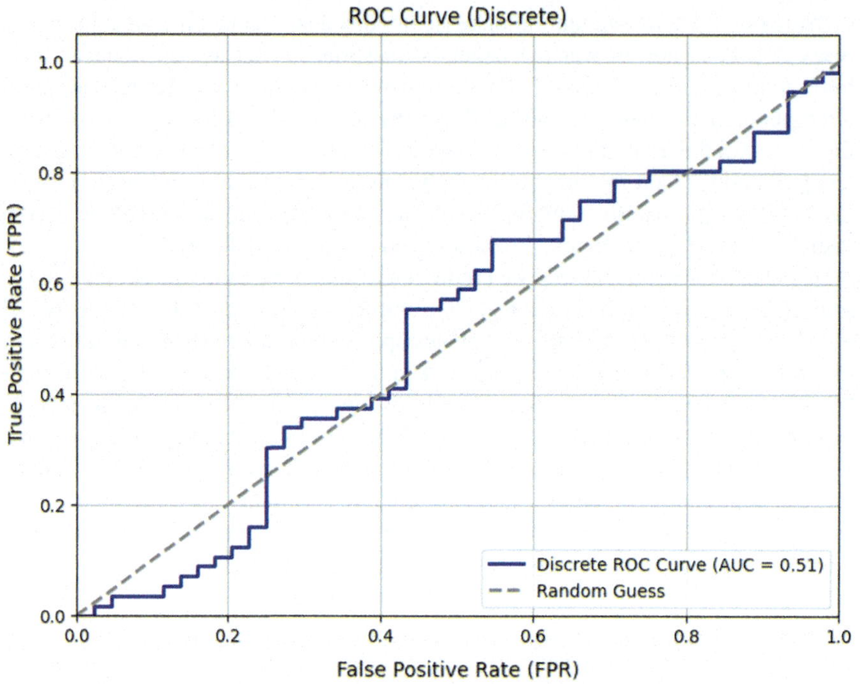


**Fig. 3.5** ROC curve for discrete case

After plotting the ROC curve, we must use the following rule to interpret its graph: we say that one point in the 2D space is better than another, if its location is at the northwest. In other words, the prediction is better if the TP is high and FP low, or TP and FP high at the same time. In the graph you can see that there is a dotted line that goes right in the middle of the plot. That line is the *line of no-discrimination*. If the points are above the line of no-discrimination, then the classification is good, otherwise it is poor.

Let's see how the ROC curve looks like for an excellent discrimination and for a failure (Figs. 3.7 and 3.8).

The ROC graph can be translated into numbers by the use of the *area under the ROC curve* (AUROC or AUC). The values of AUC are:

- 0.90 to 1.00—excellent discrimination
- 0.80 to 0.90—good discrimination
- 0.70 to 0.80—fair discrimination
- 0.60 to 0.70—poor discrimination
- 0.50 to 0.60—failure.

There should be noted that ROC curves can be used only if the dataset is balanced, meaning that we have the same number of samples for each decision class. If the dataset is not balanced, then we should use the alternative, the *Precision-Recall* (PR)

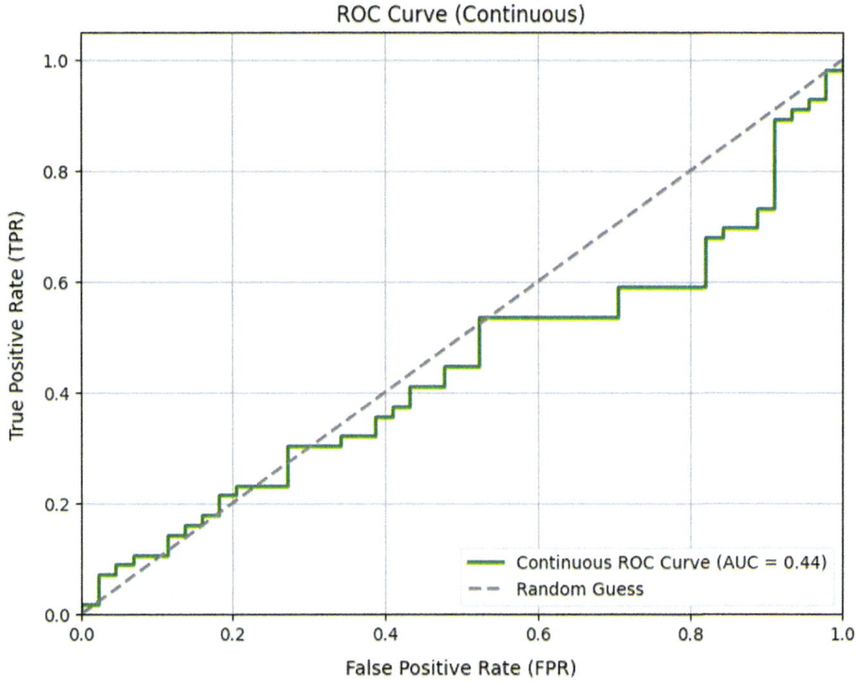


**Fig. 3.6** ROC curve for continuous case

curves. *Precision* ($P$) is computed the same as PPV, as the number of true positives over the sum of true positives and false positives.

$$Precision = PPV = \frac{TP}{TP + FP}$$

*Recall* ($R$) is computed as the number of true positives over the sum of true positives and false negatives. Recall is the same as sensitivity. Using both the recall and precision, we can compute the $F$ or $F1$ *score*.

$$F = F1 = 2 \times \frac{P \times R}{P + R}.$$

The PR curve can be summarized using the average precision $AP$, which is the weighted mean of precisions that are achieved at each threshold. We compute $AP$ as:

$$AP = \sum_i (R_i - R_{i-1}) P_i$$

where $(P_i, R_i)$ is the *operating point*, and $P_i$, and $R_i$ the $i$th threshold of the PR.

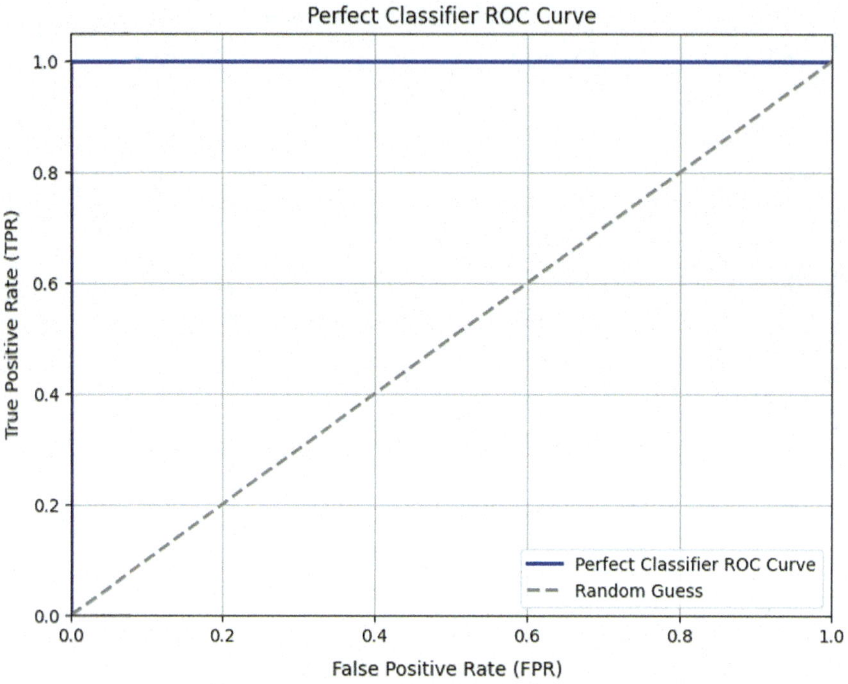


**Fig. 3.7** Perfect classifier ROC curve

More information regarding the fact that the ROC curves are too optimistic and misleading when used on unbalanced datasets can be found in [24, 25]. Nevertheless, let us state a concrete example on an imbalanced two-class decision problem. We are going to plot the two curves together with AUC for you to decide whether indeed ROC curves should be avoided when dealing with imbalanced datasets. From Figs. 3.9 and 3.10 we can see that the ROC curve is indeed too optimistic with an AUC of 0.58, whereas on PR curve's AUC is 0.16. We hope that this example shaded some light on how important it is to choose the correct measure of accuracy depending on each case.

So far, we have presented some statistical measures used to compare the results of different tests, procedures, interventions, and AI algorithms. Before continuing on this subject, let us return to the problem at hand, problem mentioned at the beginning of this chapter: over-testing, that leads to over treatment.

We are going to present an example that regards diagnosing patients with thyroid issues. The dataset which can be found at https://www.openml.org/search?type=data&sort=runs&id=40497&status=active, contains 21 attributes used to set the diagnosis. The attributes are age, sex, patient on throxine, query on throxine, patient on antithyroid medication, level of sickness, thyroid surgery, I131 treatment, query on hypothyroid, query on hyperthyroid, goitre, tumor, hypopituitary, psych, TSH,

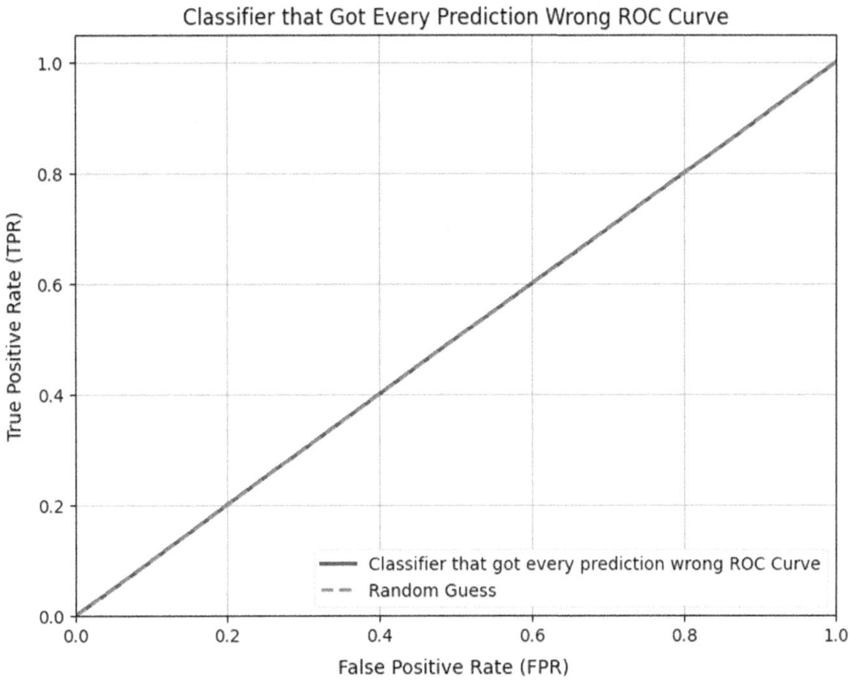


**Fig. 3.8** ROC curve failure

T31, TT4, T4U1, FTI1, and referral source. The diagnosis can be either normal, hyperthyroidism, or hypothyroidism.

We were interested in seeing whether in order to diagnose the thyroid, we needed all 21 attributes. Therefore, we have applied a combo of AI algorithms, *Random Forests* (RF) and *Artificial Neural Networks* (ANN) , to clear things out, [26]. This procedure is called *feature selection.* After applying them, 19 features were eliminated as being unnecessary, leaving only TSH (thyroid stimulating hormone) and FTI1 (free thyroxin index). This is a clear example of over testing. But just by saying that we need only 2 attributes to diagnose thyroid diseases is not enough. We need to prove that the results obtained using only the selected features versus the results obtained using all the features are the same. In this example after applying different statistical tests, we have seen that there is no significant drop in accuracy after the feature selection process. The accuracy remained approximately 96%.

When it comes to statistical tests it all revolves around the concept of *statistical significance*. The term statistical significance is connected to hypothesis testing, the Normal distribution, and the *p*-level.

Making a hypothesis implies making a statement in which we include independent and dependent variables, and assumptions are based on prior research information. Every made hypothesis must be testable, whether the test involves experiments, surveys, etc. When testing the hypothesis, we are verifying whether the results are

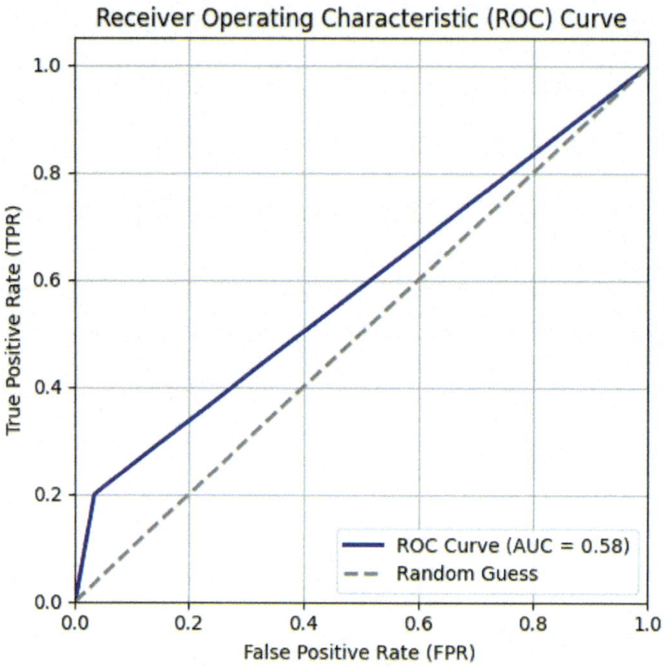


**Fig. 3.9** Example of ROC curve on an imbalanced dataset

robust, meaningful, and they have not been obtained by chance. If the results were obtained by chance, then they cannot be repeated, hence they are of no use.

Performing a statistical test involves the following three steps:

1. Establish the null hypothesis.
2. Select which is the appropriate test to verify it.
3. Perform the selected test and interpret the results. Decide whether the null hypothesis is accepted or rejected.

There are two types of tests: *parametric tests* and *non-parametric tests*. To choose between the two categories you need to know what type of distribution your data has. We use parametric tests if the data is governed by the Normal distribution. The name *parametric* refers to the mean, standard deviation, and dispersion parameters. If our data does not have a Normal distribution, then we apply the non-parametric tests, which do not use the above-mentioned parameters, nor make assumptions regarding data distribution. Each parametric test has an equivalent non-parametric test. If you still want to use parametric tests, you can transform the data from being non-Normal to approximately Normal (e.g. lognormal distribution).

To determine the level of significance we need to use another statistical concept the *p*-level or *p*-value.

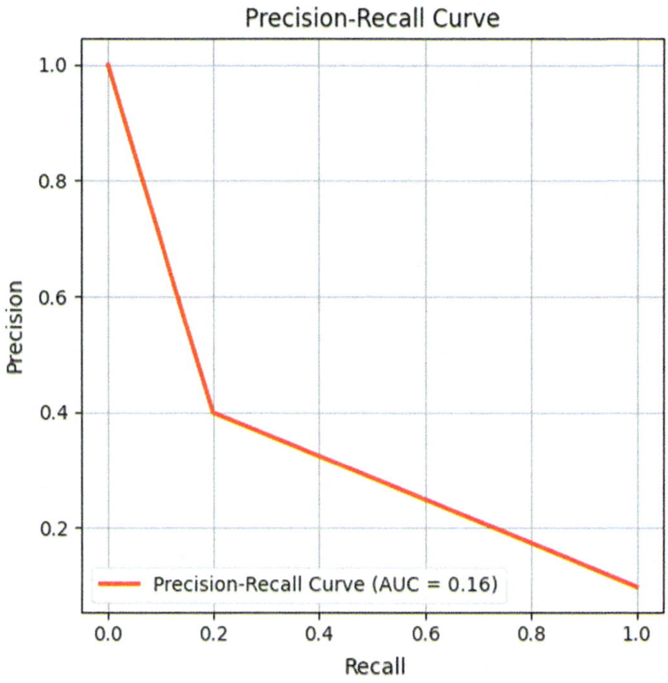


**Fig. 3.10** Example of PR curve on an imbalanced dataset

### *p-level or p-value*

Imagine that our hospital has been recently involved in clinical trials. We are testing a new drug that lowers the cholesterol. The study is performed over a group of people with high cholesterol that will try the new drug, and a control group that will be treated with an already on the market drug. The lab has to test the blood samples of patients from the two groups and report the findings. Let us presume that the new drug proves to be more efficient in lowering the cholesterol than the one that is already on the market. Obviously, just by comparing numbers we cannot prove that the clinical trial is indeed successful. We need to demonstrate that there are significant differences between the two groups, and the fact that the results obtained on the group of patients that were on the new drug were not obtained by chance. The null hypothesis in this case is: the new drug works the same as the one that is already on the market.

*p*-level was developed in 1710, by John Arbuhnot that discovered that in London more men were born than women. In fact, the chances of men being born more than women were $0.5^{82}$, which meant that it was a divine intervention, [27]. Unfortunately, it was not divine intervention, it was the social and economic status of women. Similar statistical concepts were published by Bernoulli in 1735, and by Laplace in 1778, [28, 29]. The first mention of the name *p*-value can be found in Karl Pearson's paper, in which he discussed the $\chi^2$ distribution, [30].

Thirty years later, Fisher developed the theory of the modern *p*-level, as an index that measures the significance of the evidence against the null hypothesis, [31]. In his work, Fisher has established a threshold of significance set at 0.05, and a scale to interpret the *p*-level, [32, 33].

The *p*-level cutoff value is 0.05:

- If $p$-level $\leq 0.05$, then the results are significant. There is strong evidence to reject the null hypothesis.
- If $p$-level $> 0.05$, then the results are not significant. There is no strong evidence to reject the null hypothesis. The null hypothesis will be accepted.

Two types of errors can appear when interpreting the results: type I error or $\alpha$-error, and the type II error or $\beta$-error, [34]. A type I error implies the faulty rejection of the null hypothesis, thus a false-positive result. A type II error implies the faulty acceptance of the null hypothesis, thus a false-negative result.

When presenting a test result to a patient, a test that changes their life (e.g. a test that leads to amputating a limb), we should be ready to answer all questions and most importantly to believe in the result. Have you wondered what would happen if the level of significance would be 0.0499? or 0.0501? Should we really trust the cut-off values set in the 1950s? The answer is YES. Why? Because of power analysis.

***Power analysis***

The statistical concept of *statistical power* was first introduced in the 1930s by Neyman and Pearson, [34, 35]. But as life happens, some great things are not appreciated if someone makes an opposite opinion, and that someone is an important person. Sir Roland Fisher made strong opposition, so the term vanished. In the 1960s, due to Jacob Cohen, the term was brought to life and became a powerful tool in statistics, [36].

The power of a statistical test is equal to the probability that the test's result applied on a random sample of a population is the same if applied on the whole population. The statistical power of a test is the probability of rejecting the null hypothesis when it is indeed wrong. Hence, the greater the power, the lower the chance of having a false negative (Type II error). For example, if the statistical power is 90%, then we have 10% chance of obtaining a false negative.

To know the correct sample size for an experiment, we use statistical power. In general, the larger the sample, the higher the statistical power, [37]. To compute the statistical power, we need to use *power analysis*, [38–40]. Power analysis computes the probability of finding an effect, knowing a priori that that effect exists.

We use power analysis to:

- Compute the correct sample size needed to avoid the faulty rejection of the null hypothesis.
- To compute the statistical power having a certain effect size and sample size. In some cases, we have a fixed number of patients, hence we need to know what power the statistical tests will have when applied on that fixed sample.
- To validate the research.

Power analysis is based on four statistical concepts:

- The *sample size* is the number of observations that we have in our sample.
- The *significance* is the significance level that we use in the statistical test. In general, the significance is $\alpha = 0.05$.
- The *statistical power*.
- The *effect size* is the magnitude of a result present in the population. We compute the effect size using different statistical methods such as Pearson's correlation coefficient $r$, or Cohen's $d$, etc.

The four concepts are related, that is by increasing the significance level or sample size, we can increase the statistical power, or we can easier detect the effect size. Let us see how by using different values for the significance level or different sample size we modify the power analysis, (Fig. 3.11).

The sample size is computed using *standardized difference*. The data type determines the way we compute it. Whether the data is categorical or continuous we need to ratio of the difference of interest to the interest of the standard deviation. When our data is continuous, we need to compute other parameters also:

- The power, $1 - \beta$.
- The sample data standard deviation, $sd$.
- The clinically significant difference, $\delta$.
- The significance level, $\alpha$—two sided.

The standardized difference is computed using the following formula:

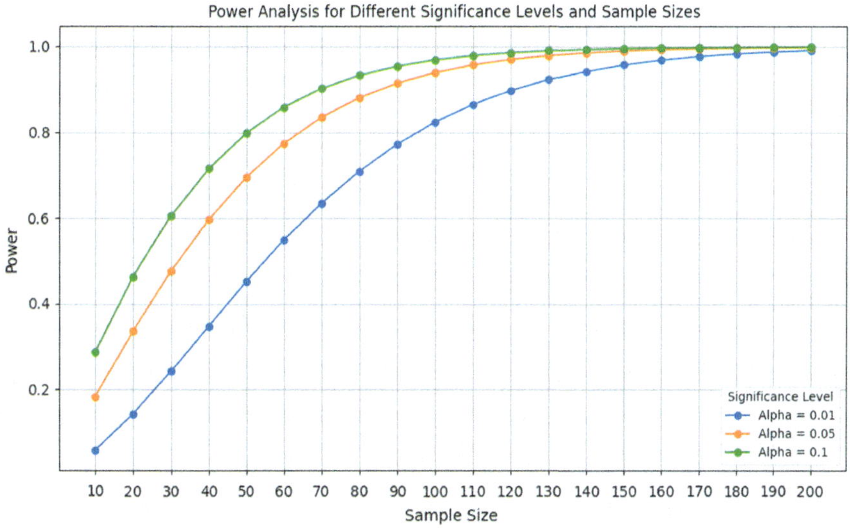


**Fig. 3.11** Power analysis for different significance levels and sample sizes

$$standardizeddifference = \frac{\delta}{sd}.$$

We cannot compute the sample size if the data is not governed by the Normal distribution or is approximately Normal. There are several ways to verify this. The first, and the easiest is to plot our data and see whether the distribution has a bell form. But this is not very accurate. For a more accurate demonstration we need to use statistical tests. So, let's start presenting the most used normality tests:

***Kolmogorov–Smirnov Goodness of Fit test (K–S)***

To verify whether our data is governed by the Normal distribution or not, we need to compare it with the Normal distribution. Hence the two hypotheses for this test are:

$H_0$ : the two samples have the same distribution. Our sample is governed by the Normal distribution.

$H_1$ : the two samples do not have the same distribution. Our sample is not governed by the Normal distribution.

The steps of applying the K–S test are, [41]:

1. Shape the sample's data empirical distribution function.
2. Select the distribution you are comparing with. In our case the Normal distribution.
3. Plot both distributions in the same graph.
4. Compute the maximum vertical distance amid the two graphs.
5. After computing the test statistic value, use the K–S table (Table A3 in the Appendix) to determine the critical value.

The K–S test computes the maximum distance between the parent distribution (Normal distribution in our case), denoted by $F_0(x)$ and our sample empirical distribution $F_{data}(x)$. The test statistic is computed using a two-tailed test:

$$D = \sup\{|F_0(x) - F_{data}(x)|, x\}.$$

If we wish to use the one-tailed test, then the absolute values are disregarded. If the computed value $D$ is greater than the critical value found in the corresponding table, table A3, then the null hypothesis is rejected.

When you decide whether the K–S test is appropriate in a certain case, you must consider the following, [42, 43]:

- The test is distribution free, which means you don't have to check the sample's distribution before performing the test.
- Computing the $D$ statistics is easy.
- The sample size doesn't need to be large.
- You need to know the sample's mean and standard deviation.
- Discrete distributions are not acceptable.
- The test has a high sensitivity at the center of the distribution, and low at the tails.

The disadvantage of having a low sensitivity at the tails has been corrected by the *Lilliefors* test. This test can be used even if we don't know the mean or the standard deviation of our data.

***Lilliefors test***

The working hypothesis for this test are the same as for the K–S test. The computation is a little bit trickier, but not too complicated. Technically, we need to compute the $z$-score for all the observations in the sample data, [44]. Below, we present the steps used to perform the Lilliefors test:

1. Compute the z-score for all the observations in the data samples:

$$z_i = \frac{X_i - \overline{X}}{sd}, i = 1, 2, \ldots, n,$$

   where:

   - $sd$ is the data sample's standard deviation
   - $X_i$ is the current observation.
   - $\overline{X}$ is the sample's mean.

2. Compute the statistic $D = max\{D^+, D^-\}$ where:

$$D^+ = \max\left\{\frac{i}{n} - f(z_i), 1 \le i \le n\right\},$$

$$D^- = \max\left\{f(z_i) - \frac{i-1}{n}, 1 \le i \le n\right\}.$$

   having the cumulative distribution function:

$$f(x) = \frac{1}{\sqrt{2\pi}} \int_{-\infty}^{x} e^{-\frac{t^2}{2}} dt.$$

3. If the computed $D$ statistics has a greater value than the critical value found in table A4 from the Appendix, then we reject the null hypothesis.

When you choose to apply the Lilliefors test to verify the normality of a sample data, please keep in mind that it has a low power.

Now, let's apply the K–S and Lilliefors test on a practical example. In Table 3.1, we have simulated the cholesterol values of 30 patients, and their corresponding $D^+$ and $D^-$.

From table 3.1, we can see that $D^+ = 0.0833$ and $D^- = 0.0900$, implying a small effect size $D = 0.0900$, thus a small magnitude between our sample distribution and the Normal distribution. The $p$-value for this statistic is 0.773, hence we accept the null hypothesis $H_0$.

**Table 3.1** Computation for K–S and Lilliefors

| Patient no | Cholesterol value | $D^+$ | $D^-$ |
|---|---|---|---|
| 1 | 169.34 | 0.00569666 | 0.0276367 |
| 2 | 178.75 | 0.0228124 | 0.010521 |
| 3 | 193.76 | 0.0152878 | 0.0180455 |
| 4 | 194.38 | 0.0464624 | – 0.013129 |
| 5 | 207.45 | 0.0243198 | 0.00901358 |
| 6 | 214.36 | 0.0202398 | 0.0130936 |
| 7 | 219.6 | 0.0214302 | 0.0119032 |
| 8 | 234.97 | – 0.0567088 | **0.0900422** |
| 9 | 236.89 | – 0.0388463 | 0.0721796 |
| 10 | 237.78 | – 0.0127796 | 0.046113 |
| 11 | 241.53 | – 0.0106625 | 0.0439958 |
| 12 | 241.71 | 0.0211502 | 0.0121832 |
| 13 | 241.83 | 0.0534687 | – 0.0201354 |
| 14 | 250.42 | 0.0124234 | 0.0209099 |
| 15 | 253.29 | 0.0204051 | 0.0129283 |
| 16 | 253.29 | 0.0537384 | – 0.0204051 |
| 17 | 253.71 | **0.0833523** | – 0.050019 |
| 18 | 258.09 | 0.0778716 | – 0.0445383 |
| 19 | 268.38 | 0.0214763 | 0.011857 |
| 20 | 270.55 | 0.0364662 | – 0.00313285 |
| 21 | 277.1 | 0.0163466 | 0.0169867 |
| 22 | 280.71 | 0.0217003 | 0.011633 |
| 23 | 283.78 | 0.0321983 | 0.00113501 |
| 24 | 289.84 | 0.0233136 | 0.0100197 |
| 25 | 292.13 | 0.0417467 | – 0.00841338 |
| 26 | 297.38 | 0.0432135 | – 0.0098802 |
| 27 | 303.37 | 0.0441823 | – 0.010849 |
| 28 | 338.28 | – 0.033598 | 0.0669313 |
| 29 | 341.15 | – 0.00469997 | 0.0380333 |
| 30 | 343.96 | 0.0247816 | 0.00855169 |

In Fig. 3.12 we present a plot that shows the sample's distribution together with $D^+$ and $D^-$.

Recall that we have mentioned that the K–S test is a low power test, hence you should always apply it together with a more powerful test to make sure that the obtained results are indeed trustworthy. In general, K–S test is applied in tandem with the Shapiro–Wilk $W$ test, which is said to be more robust.

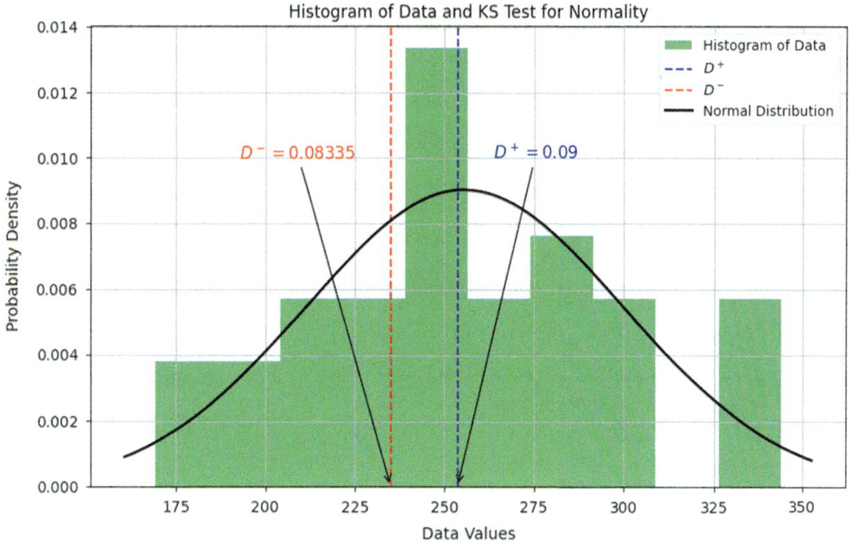


**Fig. 3.12** Histogram of Data and K–S test for normality

## Shapiro–Wilk W test

A more powerful alternative to the K–S and Lilliefors test is the Shapiro–Wilk *W* test. To decide whether we accept or reject the null hypothesis, we need to compute the *W* statistics. If the computed value is less than the critical value displayed in Table A5 from the Appendix, then we reject the null hypothesis because our data is not governed by the Normal distribution, [45–48].

The statistics *W* is computed as follows:

1. The sample data $\{x_1, x_2, \ldots, x_n\}$ is sorted in ascending order:

$$x_1 \leq x_2 \leq \cdots \leq x_n.$$

2. The $Z^2$ statistics is computed using the following formula:

$$Z^2 = \sum_{i=1}^{n} (x_i - \overline{x})^2.$$

3. The differences $d_i, i = 1, 2, \ldots, \frac{n}{2}$, for an odd $n$, or $\frac{n-1}{2}$ for an even $n$, are computed as:

$$d_1 = x_n - x_1,$$
$$d_2 = x_{n-1} - x_2,$$
$$\dots$$
$$d_i = x_{n-i+1} - x_i,$$

4. Compute $b$:

$$b = \sum_{k=1}^{i} a_k \cdot d_k,$$

where the coefficients $a_k$ can be found in Table A6 in the Appendix.

5. The W statistic is calculated as:

$$W = \frac{b^2}{Z^2}$$

After having computed the $W$ statistics, we use table AX and find the corresponding critical value.

We shall apply the Shapiro–Wilk $W$ test on the same example that involved the cholesterol values of 30 patients. After performing the above computations, we obtain $b = 238.907$ and $Z^2 = 245,815$. Hence, our $W$ statistics is 0.97, with a corresponding $p$-level $= 0.5932$, implying the acceptance of the null hypothesis. In Fig. 3.13 we have plotted the data distribution and the empirical CDF versus Normal CDF for a better visualization.

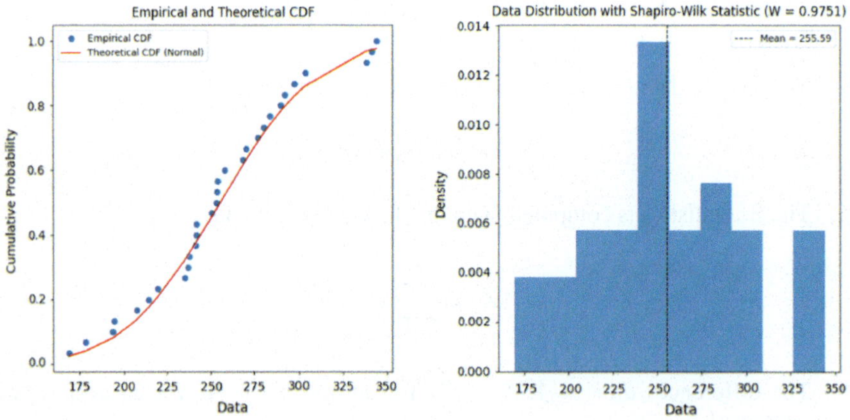


**Fig. 3.13** Data distribution with Shapiro Wilk test and empirical and Normal CDF

**Cramér-von Mises's test**

Another alternative for verifying the normality of our data sample is the Cramér-von Mises test. This test is distribution free, and it is not a powerful test. It uses the summed squared differences between the expected cumulative proportions and the observed ones to compute the test statistic. To apply it, you must follow the next steps:

1. Organize the data sample in rank order.
2. Find the data's mean and standard deviation.
3. Calculate the standardized values, denoted $z_i$.
4. Compute the $T$ statistics:

$$T = \frac{1}{12n} \cdot \sum_{i=1}^{n} \left( \frac{2i - 1}{2n} - f(z_i) \right)^2,$$

where $n$ is the sample size, and $f$ the CDF, in our case, the Normal CDF, [49, 50].

To see whether we should reject and accept the null hypothesis, we compare the obtained $T$ statistic with the tabulated value found in table A7 in the Appendix. If $T$ is greater than the tabulated value, then the null hypothesis is rejected.

When applied on our example, we obtain $T = 0.03$, with the $p$-level equaling 0.840, hence the null hypothesis is accepted.

**Anderson–Darling test**

Starting from the Cramér-von Mises test and slightly modifying it, we obtain the Anderson–Darling test. Both tests are distribution free and use distance test based on empirical CDF. Different from Cramér-von Mises, Anderson–Darling concentrates more on the observations found at the tails of the distribution, [51–55].

Since the two tests resemble, to apply the Anderson–Darling test, we need to sort the data first. Secondly, we compute the $AD$ statistic using the following formula:

$$AD = -n - \frac{1}{n} \sum_{i=1}^{n} (2i - i)\left[\ln f(x_i) + \ln(1 - f(x_{n-i+1}))\right],$$

where $n$ is the sample size, $i$ is the $i$th sample, and $f$ is the cumulative frequency function.

After computing $AD$, we look in table A8 from the Appendix to see whether its value is larger than the tabulated corresponding value. In our case $AD$ equals 0.2363, with a corresponding $p$-level of 0.7679, meaning that we accept the null hypothesis.

**D'Agostino-Pearson test**

D'Agostino-Pearson test uses moment to verify the data distribution. It computes the $K^2$ statistic that merges a test to verify the skewness and a test to verify the kurtosis, into a single test. In terms of power, D'Agostino-Pearson is powerful, but not as powerful as Shapiro–Wilk test. This is due to ignoring the existence of higher-order moments and ignoring the dependences of the moments, [56].

The skewness test is performed by converting the data's skewness into a score, as follows, [57]:

1. Compute the kth moment:

$$m_k = \sum_i \frac{(x_i - \overline{x})^k}{n}.$$

2. Compute the sample skewness:

$$b_1 = \frac{m_3}{m_2^{3/2}}.$$

3. Compute:

$$Y = \sqrt{b_1}\left(\frac{(n+1)\cdot(n+3)}{6\cdot(n+2)}\right)^{\frac{1}{2}}.$$

$$\beta_2\left(\sqrt{b_1}\right) = \frac{3\cdot\left(n^2+27n-70\right)\cdot(n+1)\cdot(n+3)}{(n-2)\cdot(n+5)\cdot(n+7)\cdot(n+9)}.$$

$$W^2 = -1 + \left(2\cdot\left(\beta_2\cdot\left(\sqrt{b_1}\right)-1\right)\right)^{\frac{1}{2}}.$$

$$\delta = \frac{1}{\sqrt{\ln W}}.$$

$$\alpha = \left(\frac{2}{W^2-1}\right)^{\frac{1}{2}}.$$

4. Compute the $Z\left(\sqrt{b_1}\right)$ statistic:

$$Z(\sqrt{b_1}) = \delta \ln\left(\frac{Y}{\alpha} + \left(\left(\frac{Y}{\alpha}\right)^2 + 1\right)^{\frac{1}{2}}\right).$$

Next, we need to transform the kurtosis into a *p*-level as follows, [58]:

1. Compute the kth moment, $m_k$:

$$m_k = \frac{(x_i - \overline{x})^k}{n}.$$

2. Compute the sample kurtosis:

$$b_2 = \frac{m_4}{m_2^2}.$$

3. Compute:

$$E\{b_2\} = \frac{3 \cdot (n-1)}{n+1}.$$

$$\operatorname{var}\{b_2\} = \frac{24 \cdot n \cdot (n-2) \cdot (n-3)}{(n+1)^2 \cdot (n+3) \cdot (n+5)}.$$

$$x = \frac{(b_2 - \mathrm{E}\{b_2\})}{\sqrt{\operatorname{var}\{b_2\}}}.$$

$$\sqrt{\beta_1(b_2)} = \frac{6 \cdot \left(n^2 - 5n + 2\right)}{(n+7) \cdot (n+9)} \cdot \sqrt{\frac{6 \cdot (n+3) \cdot (n+5)}{n \cdot (n-2) \cdot (n-3)}}.$$

$$A = 6 + \frac{8}{\sqrt{\beta_1(b_2)}} \cdot \left( \frac{2}{\sqrt{\beta_1(b_2)}} + \sqrt{1 + \frac{4}{\beta_1(b_2)}} \right).$$

where we have denoted the expected value with $E\{\cdot\}$, and the variance with $\operatorname{var}\{\cdot\}$.

4. The test statistic $Z(b_2)$ is computed using the following formula:

$$Z(b_2) = \frac{\left( \left(1 - \frac{2}{9A}\right) - \left( \frac{1 - \frac{2}{A}}{1 + x\sqrt{\frac{2}{A-4}}} \right)^{\frac{1}{3}} \right)}{\sqrt{\frac{2}{9A}}}.$$

Interpreting the D'Agostino-Pearson test is trickier than the usual interpretation of a normality test. If the sample data is normally distributed, then the D'Agostino-Pearson has chi-square distribution with 2 degrees of freedom,$\chi^2(2).\chi^2(2)$.

When applying the D'Agostino-Pearson test on our example, we find the critical value in table A9 from the Appendix. The statistics is $K^2 = 0.23$, with a corresponding $p$-value of 0.767, implying the fact that we need to accept the null hypothesis.

### Jarque and Bera test

If the sample size of our data is too large, then we cannot rely on accurate results from the above presented tests. For instance, the Shapiro–Wilk $W$ test loses its accuracy if the sample size is over 2000. Hence, we need to use another test that behaves well when applied on large datasets. Such a test is the Jarque and Bera test, which is actually a Lagrange multiplier test. To determine if the data is governed by the Normal distribution or not, the test studies the skewness and kurtosis, [59–61]. Using the D'Agostino-Pearson test, we compute the scores for the skewness and kurtosis and then use them to compute the Jarque–Bera statistic:

$$JB = n \cdot \left( \frac{\left(\sqrt{b_1}\right)^2}{6} - \frac{(b_2 - 3)^2}{24} \right).$$

To find the critical value, we must use table A10 from the Appendix. In our case, the *JB* statistics equals 0.27, with a corresponding *p*-level of 0.871, meaning we accept the null hypothesis.

Sometimes, in real medical scenarios, the collected data sample is not always governed by the Normal distribution. In this case we have three possibilities:

- Use the lognormal distribution;
- Use a non-parametric test;
- Check whether the sample size is large enough. If it is above 30 cases, then according to the *Central Limit Theorem* and the *Large Enough Sample Condition*, the sample has an approximately normal behavior. The threshold 30 has been chosen, because if you look at the *t*-table (tables A11 (one-tailed) or A12 (two-tailed) from the Appendix), you will see that the *t*-value is approximately equal to the *z* statistics for anything above *30 degrees of freedom*.

This is the second time we mention the *z*-score (Fig. 3.14), without explaining what it is. The *z*-score is a standard score that measures the distance between a sample data point and the mean of the sample. To compute it, we count the number of standard deviations that exist between the raw score and the mean. Raw score means that we must not standardize the data. Each raw score can be below or above the mean.

The *z*-score is computed as:

$$z = \frac{X - \overline{X}}{sd}.$$

If we look at the Gaussian bell, we can see the *z*-score localized between the left and right tail, between $(-3sd, 3sd)$.

As you probably have noticed, sometimes we use one-tailed tests and other times two-tailed tests. The significance level $\alpha$ is used in hypothesis testing. If we use a one-tailed test, then $\alpha = 5\%$ is found either in the left or the right tail. If we use a two-tailed test, then $\alpha$ is divided by 2, leaving $\alpha = 2.5\%$ in each tail. The cut-off areas set by $\alpha$ represent the *rejection regions*, which means that if our test statistic value falls in one of those regions then we reject the null hypothesis. So far, so good. Things seem simple enough. The difficulty is given by choosing which test to use: the one-tailed or the two-tailed? To choose the test, we need to get back to the problem. If the problem contains words such as better, worse, larger or less it is obvious. If it doesn't, then we need to dig a little deeper into the problem's input.

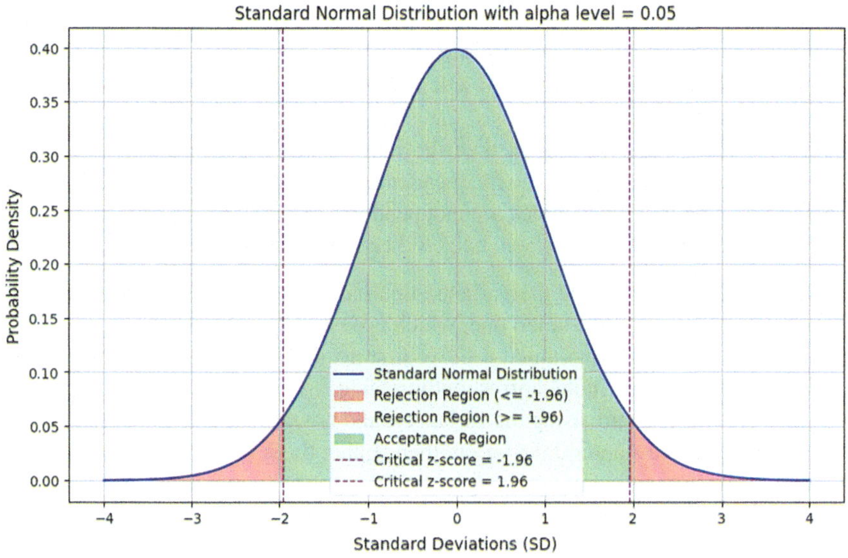


**Fig. 3.14** Representation of the $z$-score

For instance, if in our hospital there is a clinical trial that tests a drug that is used for managing multiple sclerosis, which is just as good as another one that is already on the market. The difference between the two drugs is that the new one costs less. If we were to apply a two-tailed test, then we would test whether the new drug is more or less effective than the old one. But we are not interested in the drug's effectiveness, we are interested whether if it is not less effective than the old one. So, if it is as good as the old one, it is better, because it is cheaper. Hence, we need to perform a one-tailed test, not a two-tailed one.

If we wish to compare two different tests or treatments in our laboratory department, we need to compute a *test statistic* that reveals whether the differences are significant or not. To compute a test statistic, we apply the following formula:

$$test\ statistic = \frac{observed\ value - hypothesized\ value}{standard\ error\ of\ theobserved\ value}.$$

To perform a test statistic, we first have to check whether two assumptions are true or not. The first refers to the normality of the sample data. We already have covered this subject. The second assumption refers to the *equality of variances*. The comparison statistical tests can be applied only if the variances of the two samples are approximately equal. To verify this, we can apply different statistical tests.

### *F* test

If we find ourselves in a case that involves a $F$-distribution, for instance we are performing a regression analysis, a Chow test, a Scheffe test, or we are comparing variances, we are performing the $F$ test. To compare the variances of two sample data, let's say $s_1$ and $s_2$, we need to use the $F$ statistic. Because the variances are always positive, the obtained results will always be positive.

In this case, the null hypothesis states that if the samples that are Normally distributed have equal variances, then the ratio $\frac{s_1}{s_2}$ has a $F$ distribution. A second assumption is made regarding the independence of the event. To apply the $F$ test, we need to presume that the samples are independent, [62, 63].

When you apply the $F$-test, you need to keep in mind that:

- The numerator should be the variance that has a larger value. If we do this, the test will become a right-tailed one, which is easier to compute.
- If we are applying a two-tailed test, we first need to divide $\alpha$ by 2 and then look for the critical value in the corresponding table.
- Make sure that you are using the variances, and not the standard deviations. If you have as input the standard deviation, then square them to obtain the variances.
- In the case the table you are using to find the critical value does not contain the degrees of freedom, choose the largest critical value. In this manner the Type I errors are avoided.

The $F$ test steps are:

1. Calculate the standard deviations of the two samples.
2. Square the standard deviations to obtain the variances.
3. Divide the largest one by the smallest and obtain the $F$-value.
4. Compute the degrees of freedom as the sample size minus 1. Since we work with two samples, you must have two degrees of freedom.
5. Search for the $F$-value in corresponding table A13 from the Appendix. If the computed value is larger than the tabulated one, then the null hypothesis is rejected.

Let us see how we can apply the $F$ test. We have two groups of patients, that undergo two different treatments for lowering their cholesterol levels. The first group is on a newly developed drug A, and the second group is an old drug B. The first group contains 25 patients, and the second 17. Let's see whether the two groups have equal variances or not. The data regarding the cholesterol levels is presented in Table 3.2.

Figure 3.15 presents the histograms of the sample on drug A and of the sample on drug B, plotted on the same graph, for you to spot the differences easier.

The sample on drug A has the standard deviation 79.2781, and the sample on drug B has the standard deviation 73.9333. We hypothesize that:

$H_0$: the variances of the two samples are approximately equal

$H_1$: the variances of the two sample are not approximately equal

**Table 3.2** Cholesterol levels on patients on drug A versus patient on drug B

| No. patient | Sample drug A | Sample drug B |
|---|---|---|
| 1 | 148 | 165 |
| 2 | 213 | 274 |
| 3 | 362 | 305 |
| 4 | 171 | 189 |
| 5 | 220 | 232 |
| 6 | 275 | 178 |
| 7 | 194 | 319 |
| 8 | 332 | 205 |
| 9 | 189 | 362 |
| 10 | 267 | 146 |
| 11 | 348 | 210 |
| 12 | 300 | 338 |
| 13 | 156 | 270 |
| 14 | 370 | 375 |
| 15 | 243 | 193 |
| 16 | 179 | 220 |
| 17 | 215 | 158 |
| 18 | 280 | |
| 19 | 199 | |
| 20 | 384 | |
| 21 | 152 | |
| 22 | 227 | |
| 23 | 361 | |
| 24 | 144 | |
| 25 | 340 | |

We will apply the $F$-test:

1. Compute the variance from the standard deviation.

$$s_1^2 = 6285.0233$$

$$s_2^2 = 5466.1397$$

2. Compute the $F$-value:

$$F = \frac{s_1^2}{s_2^2} = 1.1498$$

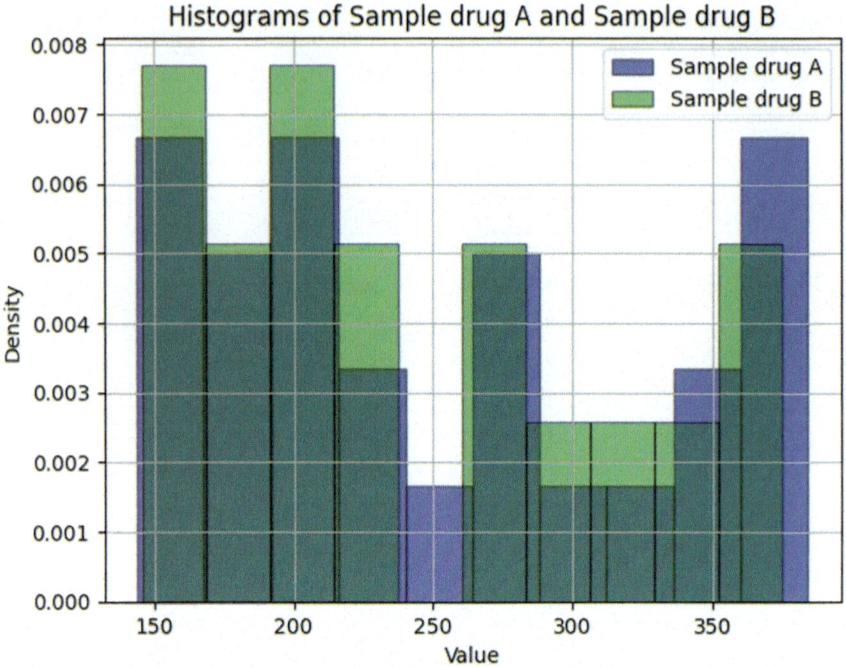


**Fig. 3.15** Histogram for the sample on drug A, and for the sample on drug B

3. Use table A13 (from the Appendix) for 24 and 16 degrees of freedom. The $p$-value is 0.7873, hence we accept the null hypothesis. The samples have approximately equal variances.

Above we have used a two-tailed test. Figure 3.16 presents the $F$-distribution plot together with the critical values for our hypothesis. The rejection and acceptance regions are highlighted on the graph for a better visualization.

If we were to use the one-tailed test, then the corresponding $p$-value for the computed $F$ statistic is 0.3936. The result does not change, the variances of the two samples are still approximately equal. What changes is the graph of the $F$-distribution. In Fig. 3.17 we have plotted the $F$-distribution with one-tail test acceptance and rejection regions.

**Bartlett's test**

Another way to verify whether two samples have approximately equal variances is by using the Bartlett's test. Note that this test can be used only if we a priori know that our samples' distributions are Normal or approximately Normal, [64, 65]. Otherwise, we should apply Levene's test. The Barlett's test statistic is compute as:

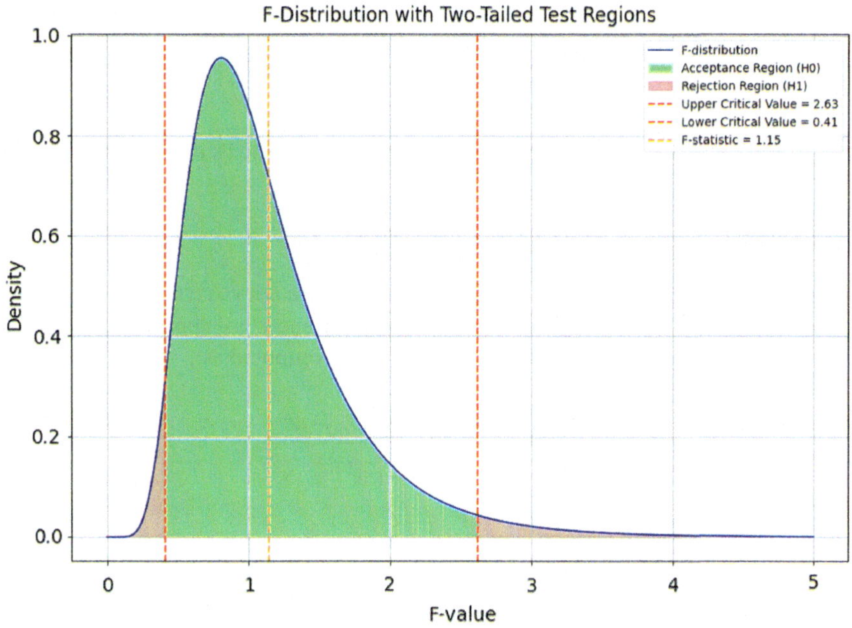


**Fig. 3.16** *F*-distribution with two-tailed hypothesis

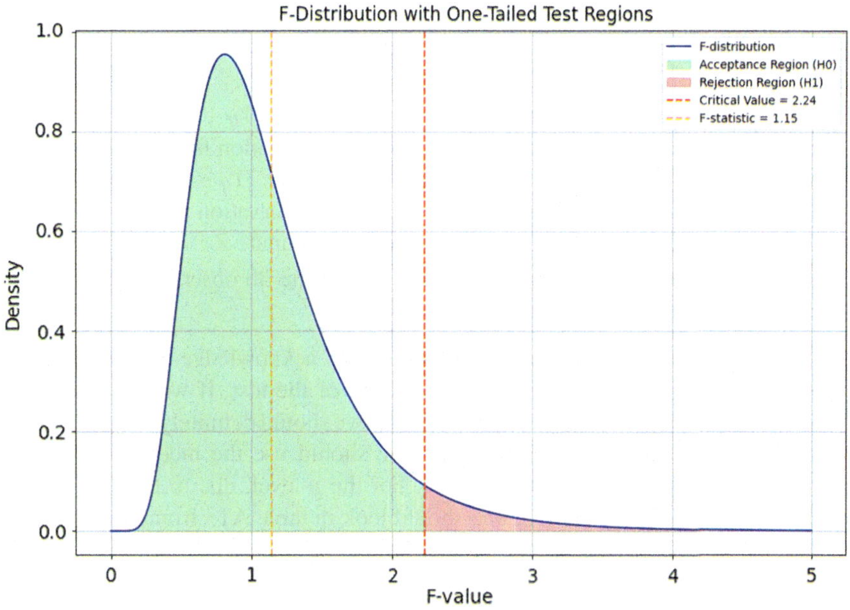


**Fig. 3.17** *F*-distribution with one-tail test

$$\chi^2 = \frac{(N-k)\ln\left(S_p^2\right) - \sum_{i=1}^{k}(n_i - 1)\ln\left(S_i^2\right)}{1 + \frac{1}{3(k-1)}\left(\sum_{i=1}^{k}\left(\frac{1}{n_i-1}\right) - \frac{1}{N-k}\right)},$$

where we denote the sample variances by $S_i$, and the variance's pooled estimate with $S_p^2 = \frac{1}{N-k}\sum_i (n_i - 1)S_i^2$.

**Levene's test**

As state above, if we a priori know that our samples are not governed by the Normal distribution, and we wish to know whether their variances are approximately equal, we should apply Levene's test. The test statistic is computed as:

$$W = \frac{N-k}{k-1} \cdot \frac{\sum_{i=1}^{k} N_i (Z_{i.} - Z_{..})^2}{\sum_{i=1}^{k}\sum_{j=1}^{N_i}(Z_{ik} - Z_{i.})^2},$$

having:

- $k$ as the number of sample groups
- $N$ the total number of observations in all the samples
- $N_i$ the number of observations in sample $i$
- $Z_{i.} = \frac{1}{N_i}\sum_{j=1}^{N_i} Z_{ij}$
- $Z_{..} = \frac{1}{N}\sum_{i=1}^{k}\sum_{j=1}^{N_i} Z_{ij}$
- If we wish $Z_{ij}$ to be the mean, then we compute $Z_{ij} = \left|Y_{ij} - \overline{Y}_i\right|$, where $\overline{Y}_i$ is the group $i$'s mean value, $Y_{ij}$ is the value of the $j$th observation from group $i$.
- If we wish $Z_{ij}$ to be the median, then we compute $Z_{ij} = \left|Y_{ij} - \widetilde{Y}_i\right|$, where $\widetilde{Y}_i$ is the group $i$'s median value, $Y_{ij}$ is the value of the $j$th observation from group $i$.
- If we wish $Z_{ij}$ to be the trimmed mean, then we compute $Z_{ij} = \left|Y_{ij} - \overline{Y}_i\right|$, where $\overline{Y}_i$ is the group $i$'s mean value, $Y_{ij}$ is the value of the $j$th observation from group $i$, for all the observations that are not outliers.

When you choose these parameters, you should acknowledge the fact that your choice will determine the power and robustness of the test. If we have a Cauchy distribution that is known to be heavy tailed, then we should definitely use the trimmed mean. If our distribution is skewed, then we should use the median, and if it is symmetric we should use the mean, [66]. For the $p$-level, due to the fact that the $W$ statistic has an $F$-distribution, one should look in table A13 from the Appendix, using $k-1$ and $N-k$ degrees of freedom.

Returning to our example, let's see what happens if we apply the Levene's test or the Bartlett's test on the two cholesterol data samples. First, we need to know whether our two samples are Normally distributed or not. For this, let us plot the two distributions together with Gauss' bell (Figs. 3.18 and 3.19). We can see from both figures that the two samples are not Normally distributed. Besides this, the sample sizes are not large enough for us to make use of the Central Limit Theorem or the Large Enough Sample Condition and assume that the sample are approximately Normally distributed. Hence, we will proceed with the Levene's test for verifying the equality of variances.

From the two graphs, we can also see that the distributions are not symmetrical, they do not have a heavy tail, hence we will apply the Levene's test using $Z_{ij}$ as the median. We obtain the $W$ statistic 0.2069, with a corresponding $p$-level of 0.6517, hence we accept the null hypothesis that states that the two samples have equal variances.

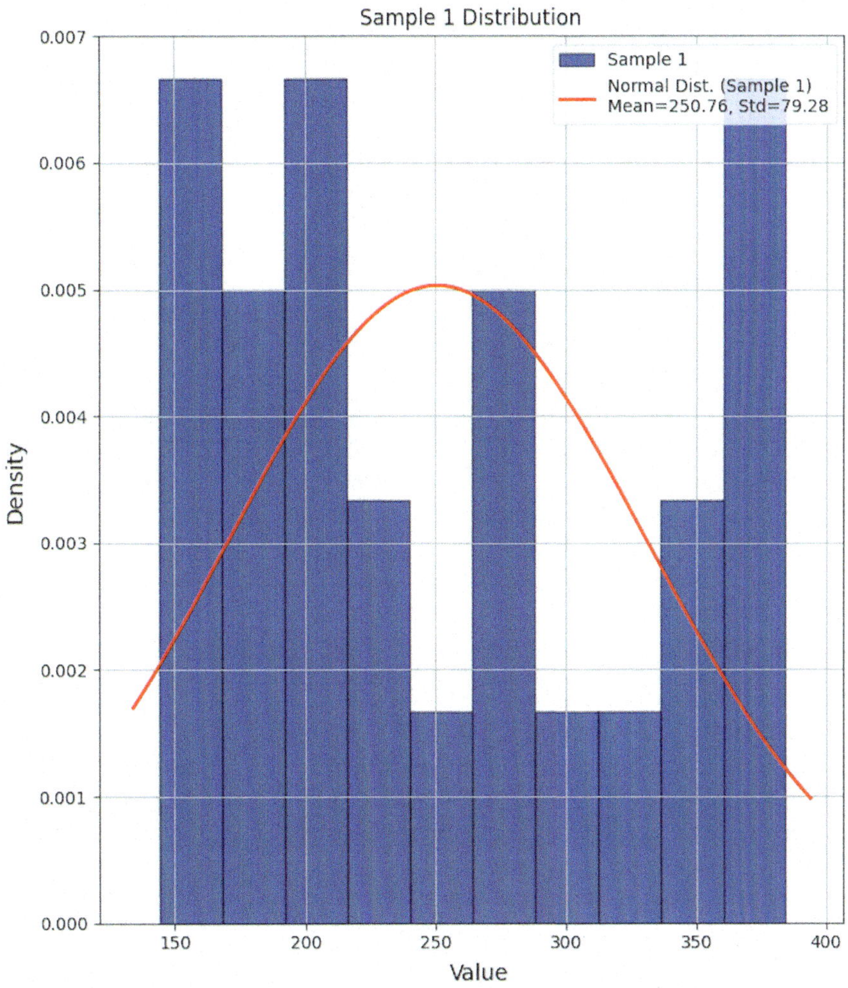


**Fig. 3.18** Distribution plot for sample on Drug A

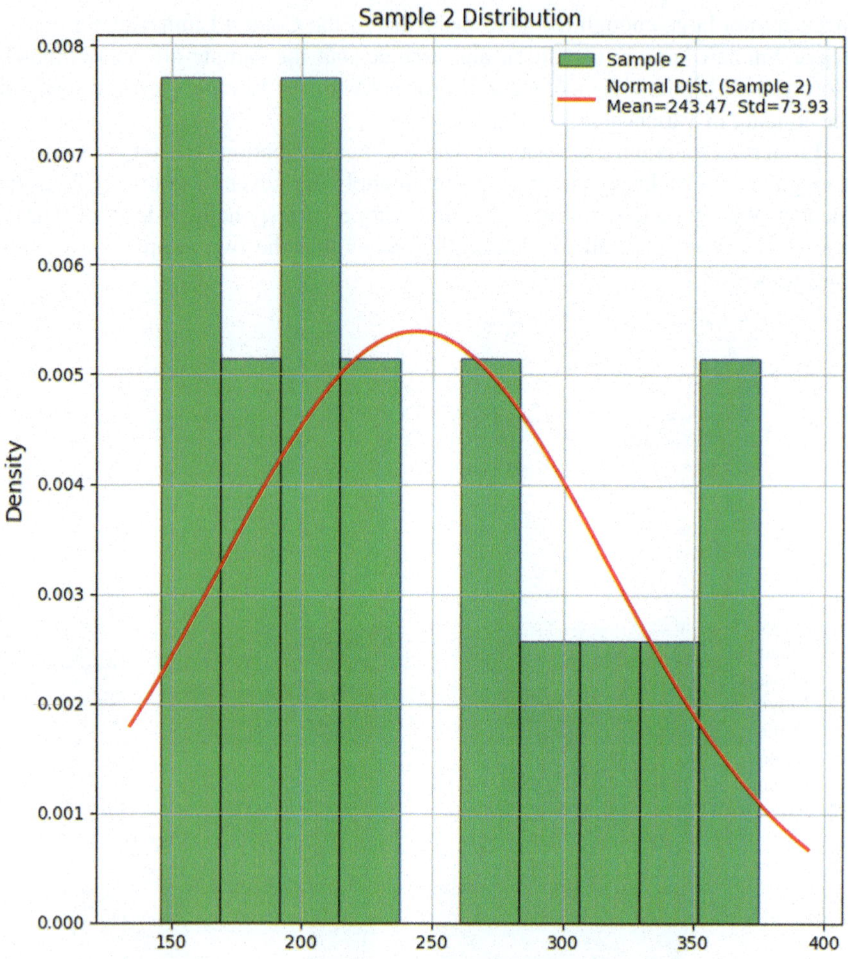


**Fig. 3.19** Distribution plot for sample on Drug B

In this chapter we have discussed different statistical measures and tests that can help us avoid over-testing and ultimately over-treating. By using different AI models, we are able to reduce the number of variables, in our case tests, that we order to set a diagnosis. More importantly, we saw how we can determine whether our tests or AI algorithms are indeed accurate or robust.

## References

1. Holtgrave, D.R., Lawler, F., Spannm, S.J.: Physicians' risk attitudes, laboratory usage, and referral decisions: the case of an academic family practice center. Med. Decis. Making **11**, 125–130 (1990)
2. Bishop, T.F., Federman, A.D., Ross, J.S.: Laboratory testing ordering at physician officies with and without on-site laboratories. J. Gen. Int. Med. **25**, 1057–1063 (2010)
3. Boland, B.J., Wollan, P.C., Silverstein, M.D.: Yield of laboratory tests for case-finding in the ambulatory general medical examination. Am. J. Med. **101**, 142–152 (1996)
4. Shapiro, M.F., Greenfield, S.: The complete blood count and leukocyte differential count. An approach to their rational application. Ann. Int. Med. **106**, 65–74 (1987)
5. Bishop, T.F., Federman, A.D., Keyhani, S.: Physicians' views on defensive medicine: a national survey. Arch. Int. Med. **170**, 1081–1083 (2010)
6. Mello, M.M., Chandra, A., Gawande, A.A., Studdert, D.M.: National costs of medical liability system. Health Aff. **29**, 1569–1577 (2010)
7. Schiff, G.D., Hasan, O., Kim, S., et al.: Diagnostic error in medicine: analysis of 583 physician-reported errors. Arch. Int. Med. **169**, 1881–1887 (2009)
8. Moynihan, R., Doust, J., Henry, D.: Preventing overdiagnosis: how to stop harming the healthy. BMJ **334**, e3502 (2012)
9. Greenberg, J., Green, J.: Over-testing: why more is not better. Am. J. Med. **127**(5), 362–363. https://doi.org/10.1016/j.amjmed.2013.10.024
10. Vrijsen, B.E.L., Naakteboren, C.A., Vos, L.M., et al.: Inappropriate laboratory testing in internal medicine inpatients: prevalence, causes and interventions. Ann. Med. Surg. **51**, 48–53 (2020)
11. Treadwell, J., McCartney, M.: Overdiagnosis and overtreatment: generalists—it's time for a grassroots revolutions. BJGP **66**(644), 116–117 (2016)
12. Moynihan, R., Barrat, A., Buchbinder, R., et al.: Australia is responding to the complex challenge of overdiagnosis. MJA **209**(8), 332 (2018)
13. Jenniskens, K., de Groot, J., Reitsma, J., et al.: Overdiagnosis across medical disciplines: a scoping review. BMJ Open **7**, e018338 (2017)
14. Pathirana, T., Clark, J., Moynihan, R.: Mapping the drivers of overdiagnosis to potential solutions. BMJ **358**, j3879 (2017)
15. Agarwal, R.: Quality-improvement measures as effective ways of preventing laboratory errors. Lab. Med. **45**, e80–e88 (2014). https://doi.org/10.1309/LMD0YIFPTOWZONAD
16. Aesif, S.W., Parenti, D.M., Lesky, L., Keiser, J.F.: A cost-effective interdisciplinary approach to microbiologic send-out test use. Arch. Pathol. Lab. Med. **139**, 194–198 (2015)
17. Bayes, T., Price, M.: An essay toward solving a problem in the doctrine of chances. By the late Rev. Mr. Bayes F.R.S. communicated by Mr. Price, in a letter to John Canton, A.M.F.R.S. Philos. Trans. R. Soc. Lond. **53**, 370–418 (1763). https://doi.org/10.1098/rstl.1793.0053
18. Pepe, M.S.: The Statistical Evaluation of Medical Tests for Classification and Prediction. Oxford University Press, Oxford (2004)
19. Zweig, M.H., Campbell, G.: Receiver-operating characteristic (ROC)—plots: a fundamental evaluation tool in clinical medicine. Clin. Chem. **39**(8), 561–577 (1993)
20. Zur, R.M., Pesce, L.L., Jiang, Y.: The effect of two priors on Bayesian estimation of "Proper" binormal ROC curves from common and degenerate datasets. Acad. Radiol. **17**(8), 969–979 (2010)
21. Roemer, V.M., Walden, R.: Sensitivity, specificity, receiver-operating characteristic (ROC) curves and likelihood ratios for electronic fetal rate monitoring using new evaluation techniques. Z. Geburtshilfe Neonatol. **214**(3), 108–118 (2010)
22. Jin, H., Lu, Y.: A non-inferiority test of areas under two parametric ROC curves. Contemp. Clin. Trials. **30**(4), 375–379
23. Lasko, T.A., Bhagwat, J.G., Zou, K.H., Ohno-Machado, L.: The use of receiver operating characteristic curves in biomedical informatics. J. Biomed. Inf. **38**(5), 404–415 (2005)

24. Davis, J., Goadrich, M.: The relationship between Precision-Recall and ROC curves. In: ICML'06 Proceedings of the 23rd International Conference on Machine Learning, Pittsburgh, PA, USA, pp. 233–240. https://doi.org/10.1145/1143844.1143874
25. Saito, T., Rehmsmeier, M.: The precision-recall is more informative than the ROC plot when evaluating binary classifiers on imbalanced datasets. PloS ONE **10**(3) (2015). https://doi.org/10.1371/journal.pone.0118432
26. Belciug, S., Gorunescu, F.: Improving performance in neural networks using feature selection based on random forests: application to automated medical diagnosis. Int. J. Inf. Tech. Knowl. **10**(1), 33–46 (2016)
27. Arbuthnot, J.: An argument for divine providence, taken from the constant regularity observed in the births of both sexes. Philos. Trans. R. Soc. Lond. **27**, 186–190 (1710)
28. Bernoulli, D.: Pieces que ont remporte le prix double de l'Academie Royale des Science Recherchers physiques et astronomiques **1734**, 3–122 (1735)
29. LaPlace, P.: Memoires de l'Academie Royale de Sciences de Paris **9**, 227–332 (1778)
30. Pearson, K.: On the criterion that a given system of deviations from the probable in the case of a correlated system of variable is such that it can be reasonable supposed to have arisen from random sampling. Lond. Edinb. Dublin Philos. Mag. J. Sci. Ser. **5**(50), 157–175 (1900)
31. Dahiru, T.: *P*-value, a true test of statistical significance? A cautionary note. Ann. Ib. Postgrad. Med. **6**(1), 21–26 (2008)
32. Bakan, D.: The test of significance in psychological research. Psychol. Bull. **66**, 423–437 (1960)
33. Biau, D.J., Jolles, B.M., Porcher, R.: *p*-value and the theory of the hypothesis testing: an explanation for new researchers. Clin. Orthop. Relat. Res. **468**(3), 885–892 (2010). https://doi.org/10.1007/s11999-009-1164-4
34. Neyman, J. Pearson, E.S.: On the use and interpretation of certain test criteria for purposes of statistical inference. Biometrica **20a**, 175–240, 263–294 (1928)
35. Neyman, J., Pearson, E.S.: On the problem of the most efficient tests of statistical hypotheses. Trans. R. Soc. Lond. Ser. A **231**, 289–337 (1933)
36. Cohen, J.: The statistical power of abnormal-social psychological research: a review. J. Abnom. Soc. Psychol. **65**(3), 145–153 (1962). https://doi.org/10.1037/h0045186
37. Altman, D.G.: Practical Statistics for Medical Research. Chapman and Hill (1991)
38. Kemal, O.: Power analysis and sample size, when and why? Turk. Arch. Otorhinolaryngol. **58**(1), 3–4 (2020). https://doi.org/10.5152/tao.2020.0330
39. Joens, S.R., Carley, S., Harrison, M.: An introduction to power and sample size estimation. BMJ Emerg. Med. J. **21**(1), 126 (2004)
40. Arkin, C.F., Wachtel, M.S.: How many patients are necessary to assess test performance? JAMA **3**, 895–900 (1990)
41. Chakravarti, I.M., Laha, R.G., Roy, J.: Handbook of Methods of Applied Statistics, vol. I, pp. 392–394. Wiley, Hoboken (1967)
42. Zhang, J., Wu, Y.: Beta approximation to the distribution of Kolmogorov-Smirnov statistics. Ann. Inst. Stat. Math. **54**(3), 577–584 (2001)
43. Dimitrova, D., Kaishev, V.K., Tan, S.: Computing the Kolmogorov-Smirnov distribution when the underlying CDF is purely discrete, mixed or continuous. J. Stat. Softw. **95**(10) (2020). 10.18637.jss.v095.i10
44. Lilliefors, H.W.: On the Kolmogorov-Smirnov test for normality with mean and variance unknown. J. Am. Stat. Assoc. **62**, 399–402 (1967)
45. Mbah, A.K., Pothong, A.: A Shapiro-Francia test compared to other normality test using expected *p*-value. J. Stat. Comput. Simul. **85**, 3002–3016 (2014). https://doi.org/10.1080/00949655.2014.947986
46. Shapiro, S.S., Wilk, M.B.: An analysis of variance test for normality (concrete samples). Biometrika **52**, 591–611 (1965)
47. Shapiro, S.S., Francia, R.S.: An approximate analysis of variance test for normality. J. Am. Stat. Assoc. **67**, 25–216 (1972)
48. Royston, P.: A pocket-calculator algorithm for the Shapiro-Francia test for normality—an application to medicine. Stat. Med. **12**(2), 181–184 (1993). https://doi.org/10.1002/sum.4780120209

49. Cramer, H.: On the composition of elementary errors. Skandinavisk Aktuarietidskrift **11**(13–74), 141–180 (1928)
50. von Mises, R.:Wahrscheinlichkeitsrechnung und Ihre Anwendung in der Statistik und Theoretischen Physik, Julius Springer (1931)
51. Anderson, T.W., Darling, D.A.: Asymptotic theory of certain goodness of fit criteria based on stochastic processes. Ann. Math. Stat. **23**, 193–212 (1952)
52. Anderson, T.W., Darling, D.A.: A test of goodness of fit. J. Am. Stat. Assoc. **49**, 765–769 (1954)
53. Durbin, J., Knott, M., Taylor, C.C.: Components of Cramer-von Mises statistics, II. J. R. Stat. Soc. Ser. B **37**, 216–237 (1975)
54. Stephens, M.A.: EDF statistics for goodness of fit and some comparisons. J. Am. Stat. Assoc. **69**, 730–737 (1974)
55. Stephens, M.A.: Tests based on EDF statistics. In: D'Agostino, R.B., Stephens, M.A. (eds.) Goodness-of-Fit Techniques. Marcel Dekker, New York (1986)
56. D'Agostino, R.B., Pearson, E.S.: Tests for departure from normality. Biometrika **60**, 613–622 (1973)
57. D'Agostino, R.B.: Transformation to normality of the null distribution of g1. Biometrika **57**(3), 679–681 (1970)
58. Anscome, F.J., Glynn, W.J.: Distribution of the kurtosis statistic b2 for normal samples. Biometrika **70**(1), 227–234 (1983)
59. Jarque, C.M., Bera, A.K.: Efficient tests for normality, homoscedasticity and serial independence of regression residuals. Econ. Lett. **6**(3), 255–259 (1980)
60. Jarque, C.M., Bera, A.K.: Efficient tests for normality, homoscedasticity and serial independence of regression residuals: Monte Carol evidence. Econ. Lett. **7**(4), 313–318 (1981)
61. Jarque, C.M., Bera, A.K.: A test for normality of observations and regression residuals. Int. Stat. Rev. **55**(2), 163–172 (1987)
62. Archdeacon, T.: Correlation and Regression Analysis: A Historian's Guide. University of Wisconsin Press (1994)
63. Sawilwosky, S.: Fermat, Schubert, Eistein, and Behrens-Fisher: the probable difference between two means when $\sigma_1^2 neq \sigma_2^2$. J. Mod. Appl. Stat. Methods **1**(2), 461–472 (2022)
64. Bartlett, M.S.: Properties of sufficiency and statistical tests. Proc. R. Stat. Soc. A **160**, 268–282 (1937)
65. Snedecor, G.W., Cochran, W.G.: Statistical Methods, 8th edn. Iowa State University Press (1989)
66. Brown, M.B., Forysthe, A.B.: Robust tests for equality of variances. A.B.J. Am. Stat. Assoc. **69**, 364–367 (1974)

# Chapter 4
# Intelligent Inpatient and Outpatient Management

**Abstract** The aim of this chapter is to see how we could make the inpatient and outpatient management intelligent, using operation research algorithms. We will discuss different queuing models, exemplify them through real life scenarios, and answer questions on how to optimize the inpatient and outpatient management.

Sometimes it seems like the most part of our lives we wait. We wait in line to buy food, to get into the theatre, to buy fuel, to board a plane, etc. Sometimes we wait patiently, other times we are annoyed, and in some cases, we even leave, letting things unresolved. The hardest part is waiting for a doctor. While sitting in the waiting room, you start to imagine the worst. Knowing is better than not knowing. Luckily, queuing models have been built.

There are two types of waiting lines: waiting lines that we can see, because there are right in front of us (e.g. waiting at the pharmacy, waiting at the doctor's office etc.), and waiting lines that we cannot see (e.g. failing to talk to someone on the phone because the line is busy, a surgery that is postpone because the O.R. is occupied due to unforeseen complications in another surgery, etc.).

Queuing models are stochastic models. Queuing models are fruitful because they are modeled using the stochastic processes discussed in Chap. 2. We can optimize queuing models if we understand what factors influence our system. Potential factors are the arrival rate of patients, the number of service providers, and the service time. Think of the service providers in terms of medical personnel, and the service time in terms of different procedures or treatments.

Inpatient means, in the most basic sense of the term, a person that has been admitted to the hospital to spend at least one night. Inpatients are people that need to be monitorized closely. Some inpatient care examples are: routine and complex surgeries, serious medical issues that need to be monitored, childbirth, rehabilitation services after strokes, severe injuries, substance misuse, psychiatric illnesses, etc. An inpatient has contact with large groups of healthcare providers. While being admitted to the hospital, a patient interacts with doctors, nurses, lab technicians, pharmacists, physical therapists, etc.

 
S. Belciug, *The Hospital Manager's Guide to Artificial Intelligence*, Intelligent Systems Reference Library 263, https://doi.org/10.1007/978-3-031-80314-7_4

**Fig. 4.1** An example of single server queue

Outpatient care represents the ambulatory care, or in other terms any treatment or service provided to the patient that does not require hospitalization overnight. Examples of outpatient care services include but are not limited to minor surgeries, colonoscopies, endoscopies, mammograms, CTs, MRIs, X-rays, consultations or follow-ups, routine physical exams, medical problems that are treated in the ER, chemotherapy or radiation treatment, etc.

Queuing theory was developed by Agner Krarup Erlang, a Danish mathematician, statistician, and engineer, to describe the Copenhagen telephone exchange, [1]. Queuing theory is part of the operational research domain.

In general, queuing models have been applied in industry. We will try to prove that queuing models can be applied in hospital management also. Let us start with presenting some terms of queuing models. In a hospital or ambulatory, we have patients that expect a *service* (e.g. diagnosis, blood work, imaging tests, treatment, surgery, etc.). If a patient arrives at a certain hospital department (e.g. radiology), where a *server* is free, then that patient can be served by the server instantly. The server in this case can be a CT or an MRI machine. If no server is free, all the CT and MRI machines being occupied, then the patient will have to wait, which means that a queue is being created, [2].

The system can have a single *channel* that corresponds to a single server (Fig. 4.1), or multiple channels that correspond to multiple servers (Fig. 4.2). A multiple server queue is a queueing model that has only one waiting line that ends with multiple servers (i.e. the patient does not care which doctor consults him/her/they). In other cases, we can have multiple servers with multiple waiting lines (i.e. the patient wishes to be consulted by a certain doctor) (Fig. 4.3).

In some cases, such as in congestion theory, the term service time is replaced with holding costs. In our case, if a patient is admitted to the ICU after a surgery, then the service time/holding cost is represented by the number of days the patient stays in the ICU, hence the number of days a bed is occupied.

When describing a queuing model we use the following concepts, [3]:

- *Population*. The population can be *finite* or *infinite*. When discussing patients, the queuing models deal with infinite population, because a patient from an infinite population can arrive at the hospital at any time. When discussing a machine such as an MRI machine, a CT scanner, etc., then we are dealing with a finite

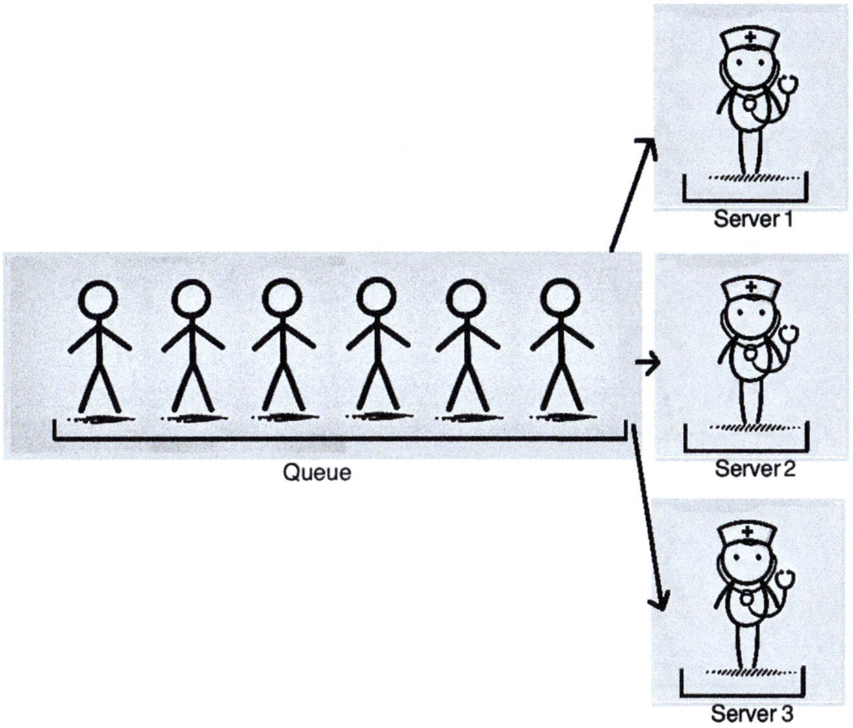


**Fig. 4.2** An example of a multiple server queue and a single waiting line

population, since we cannot have an infinite amount of equipment in a healthcare facility.

- *Service* is the way the patients will be served: admitted, diagnosed, consulted, operated on, etc.
- *Server* is the resource providing the service: the doctor, the nurse, the MRI, CT machines, the bed, etc.
- *Interarrival Time Distribution*. Patients arrive at the hospital or ambulatory expecting to be diagnosed and treated. In general, they come one by one, or *en masse* (multiple casualties in an accident). We can describe the interarrival pattern of patients, if we take into account the interarrival time distributions of their arrivals. The following interarrival times distributions can be considered:
  - *Poisson arrivals* is denoted in queuing theory by $M$. The $M$ notation is used to show that the process has Markovian (memoryless) property. If the patients arrive at a hospital following a Poisson arrival pattern, then the interarrival time distribution follows the exponential distribution:

$$F(t) = 1 - e^{-\lambda t}, t \geq 0,$$

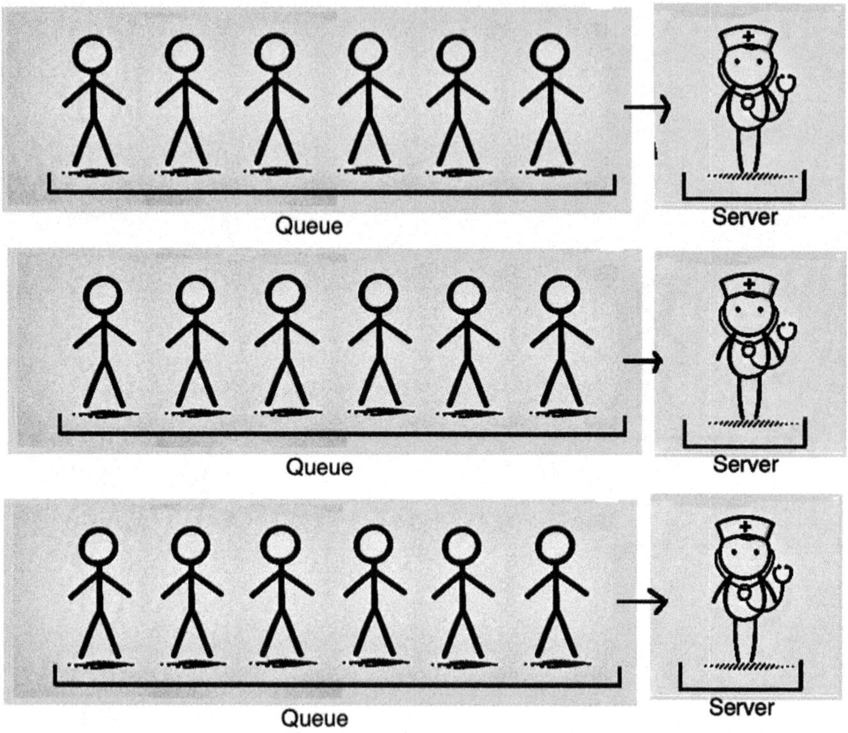


**Fig. 4.3** An example of a multiple server queue and multiple waiting lines

where $\lambda$ denotes the parameter of a Poisson arrival and it is known as the arrival rate, and $\frac{1}{\lambda}$ is the mean arrival time for each patient. Recall that a Poisson process describes random arrivals.

- *k-Erlang Distribution* is denoted by $E_k$. The interarrival time distribution of patients follows a *k-Erlang distribution*, and is computed as follows:

$$F(t) = \int_{0}^{t} \frac{k\lambda(k\lambda x)^{k-1} e^{-k\lambda x}}{(k-1)!} dx$$

where $t \geq 0$, $\lambda > 0$, and $k$ is a positive integer. The $k$-Erlang distribution is a gamma distribution, $T \sim GAM\,(k\lambda, k)$, with the mean-$E()$ and variance- $Var()$ computed as:

$$E(T) = \frac{1}{\lambda},$$

$$Var(T) = \frac{1}{k\lambda^2}.$$

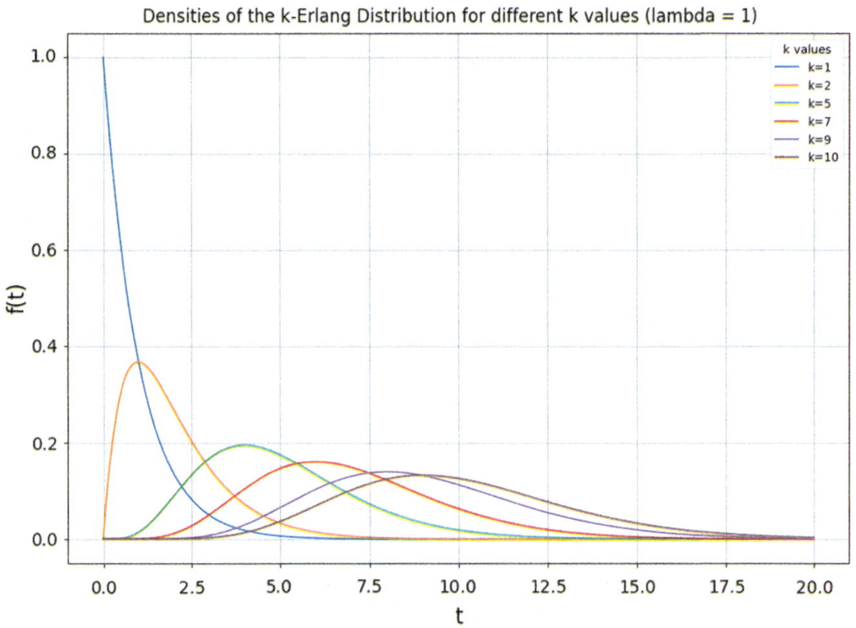


**Fig. 4.4** *k*-Erlang distribution for different values of *k*

In Fig. 4.4 we present how the density of the $k$-Erlang distribution changes its shape for $\lambda = 1$, and different values for $k$. You can see from Fig. 4.4 that if $k = 1$, then we are dealing with the exponential distribution. As $k$ increases it produces a degenerate distribution at $t = \frac{1}{\lambda}$. We present more details on *degenerate distribution* below.

- *Degenerate Distribution* is denoted by $D$. The interarrival time distribution is pretty basic, patients arriving methodically at $t = \frac{1}{\lambda}$.

$$F(t) = \begin{cases} 0, t < \frac{1}{\lambda} \\ 1, t \geq \frac{1}{\lambda} \end{cases},$$

with the mean and variance computed as:

$$E(T) = \frac{1}{\lambda},$$

$$Var(T) = 0$$

- *General Distribution* denoted by $G$. We use this distribution when we believe that the patients arrive independently and identically distributed following a random distribution $F(t), t > 0$.

All the examples from this book will follow a Poisson interarrival time distribution.

- *Service Time Distribution*. The same distributions that we have discussed for the interarrival time distribution can be applied for the service time. We denote the service rate with $\mu$. If the service time distribution is a Poisson distribution, then it follows an exponential distribution.
- *Maximum Queuing System Capacity*. Here we have two choices: either we say that the system has an infinite capacity, meaning that all the patients that arrive are allowed to wait as long as they like, regardless of how the queue is when they arrive, or we say that the system has a finite capacity, meaning that a patient can arrive when the maximum capacity of the system has been reached, hence the patient is turned away. For instance, an infinite capacity system is the ER department, but a hospital department with allocated beds has a finite capacity (i.e. the number of beds). If there is no queuing space in a finite capacity system, then we have a "loss" system, in which each turned away patient represents a loss to our system, and possibly a penalty cost.
- *Number of Service Channels*. This subject has already been discussed (see Figs. 4.3 and 4.4). We can have a simple queuing model that has a single service channel with one server, or a multiple server queueing model, that has multiple servers, but only one waiting line, or multiple waiting lines.
- *Queuing discipline*. Patients wait for service. Here we are not going to triage them according to different priority keys. The order in which the patients are served represents the *queue discipline*. In general, the rule of *first come, first served (FCFS)* is applied. In this book, we are going to use this type of rule. Nevertheless, we shall mention other queue disciplines, such as *last come, first served*; *random selection of service*; *shortest selection for service* and *service according to priority*.
- *Queuing behavior* describes the way patients act while waiting in line. We have different behaviors:

    - The *balk behavior* is when the patient leaves the queue when she/he/they sees it is too long.
    - The *renege behavior* is when the patient leaves the queue because the line is moving too slow, and it seems it is a waste of time. This applies when the patient's condition allows him/her/they to leave without his/hers/their life to be threatened.
    - The *jockey behavior* is noticed when we have multiple servers and multiple queues. The jockey patient switches between the lines, thinking that way will get him/her/they served faster.

A queuing model has the following notation, developed by David George Kendall: $A/B/c/K/m/Z$, where

- $A$ represents the interarrival time distribution.
- $B$ represent the service time distribution.
- $c$ represents the number of servers.
- $K$ represents the waiting system's capacity.

**Fig. 4.5** $M/M/1/FCFS$ model

- $m$ represents the population size.
- $Z$ represents the queue discipline.

The maximum queuing system capacity is computed as the sum of the maximum number of patients waiting in line plus the number of service channels. In this book, we will study the queueing models that have the form $M/M$, that is Poisson arrival and exponential service, with different other components.

As a hospital manager, we can use queuing theory to help us manage the resources. Here are some questions that we would like to have the answer to:

- What are the chances (probability) of having no patients in the system?
- What is the average number of patients that are waiting?
- What is the average number of patients in the whole system (patients that are waiting plus patients that are being treated in the meantime)?
- How much time does a patient have to wait to be seen by a doctor? Or to have an imaging test performed? Etc.
- How much time does a patient spend in the system (waiting and service time)?
- What are the chances of a patient that has just arrived waiting to be seen by a doctor, etc.?
- What are the chances that at a certain moment to have $n$ patients in the system?

Next, we are going to present some mathematical queueing models that can answer these questions. To get the correct answer first you have to determine which type of waiting line you are dealing with. We are going to model different situations, with no triage. Having these examples you can model any other situation.

***Single Server Queuing Models $M/M/1/FCFS$***

We are modeling a single channel waiting line. We have a receptionist at the admission desk that processes the form filled out by the patient. Once a patient is recorded in the system, the receptionist starts processing the forms of the next patient. Figure 4.5 presents a diagram of this system.

We mentioned that the arrival pattern follows a Poisson distribution. Hence the probability of having $k$ arrivals in a given period of time is computed using the following formula:

$$p_x(k) = P(X = k) = \frac{e^{-\lambda} \cdot \lambda^k}{k!}, k = 0, 1, 2, \ldots$$

where

- $k$ is the number of arrivals in a given period of time
- $\lambda$ is the average number of arrivals in the given time period.

This is the theory, but how do we apply it in practice? First of all, we need to record how many arrivals are in a given period of time, whether we are discussing hours, days, weeks, etc. Having these records, we can determine whether their distribution is Poisson or not.

Next, we need to know how long it takes to provide the healthcare service. The service time is case dependent. It has been proven that the exponential distribution approximates well service times. If we do not want patients to walk away, we need to tell them how long they have to wait to be served. Therefore, if we wish to know what the chances for a patient are to wait less, or at least equal to a given time $t$, we compute:

$$P(servicetime \leq t) = 1 - e^{-\mu t},$$

where we denote $\mu$ as the average number of patients that can be served in time $t$.

To answer the above-mentioned questions, we must make sure that our system is in a *steady-state*. This implies that the computations will be made without taking into consideration the starting and ending hours.

Consider a hospital ER department where patients arrived, but we have only one doctor on call. What are the key performance measures for this system? Let us answer the questions one by one. What do we know? We know that the arrival rate is 3 patients per hour, and the service rate is 4 patients per hour, hence $\lambda = 3$, $\mu = 4$.

- What are the chances (probability) of having no patients in the system?

$$P_0 = 1 - \frac{\lambda}{\mu} = 0.25.$$

  There are 25% chances that the E.R. is empty.
- What is the average number of patients that are waiting?

$$L_q = \frac{\lambda^2}{\mu(\mu - \lambda)} = \frac{3^2}{4(4-3)} = \frac{9}{4} = 2.25$$

  The average number of patients in the queue is 2.25 patients.
- What is the average number of patients in the whole system (patients that are waiting plus patients that are being treated in the meantime)?

$$L = L_q + \frac{\lambda}{\mu} = 2.25 + 0.75 = 3$$

The average number of patients in the system is 3.

- How much time does a patient have to wait to be seen by a doctor?? Etc.?

$$W_q = \frac{L_q}{\lambda} = \frac{2.25}{3} = 0.75$$

The average time that a patient has to wait to be seen by the doctor is 0.75 hours, that is 45 min.

- How much time does a patient spend in the system (waiting and service time)?

$$W = W_q + \frac{1}{\mu} = 0.75 + \frac{1}{4} = 1$$

The average time that a patient spends in the ER is 1 h.

- What are the chances of a patient that has just arrived to wait to be seen by a doctor, etc.?

$$P_w = \frac{\lambda}{\mu} = \frac{3}{4} = 0.75$$

There are 75% chances for a patient to wait to be seen by a doctor.

- What are the chances that at a certain moment to have $n$ patients in the system?

$$P_n = \left(1 - \frac{\lambda}{\mu}\right) \cdot \left(\frac{\lambda}{\mu}\right)^n$$

The chances of having 0 patients in the system? 25%.
The chances of having 1 patient in the system? $(1 - 0.75) \cdot 0.75^1 = 0.18$.
The chances of having 2 patients in the system? 0.14.
The chances of having 3 patients in the system? 0.10.
The computations can be done if and only if the mean service is greater than the mean arrival rate.

***M/M/k/FCFS* model**

The *M/M/k/FCFS* model is used when we have multiple servers, *k*, with one waiting line. All the servers have the same service rate. Let us expand the above example by placing two or more receptionists at the admission desk (Fig. 4.6).

The operating characteristics of this type of model can be computed, if we a priori verify whether the following assumptions are true or not:

- The system has two or more servers
- The interarrival time follows a Poisson distribution
- The service time follows an exponential distribution
- The servers have the same mean service rate
- As the patients come, they form one waiting line. When a server frees up, a patient moves onto that server.

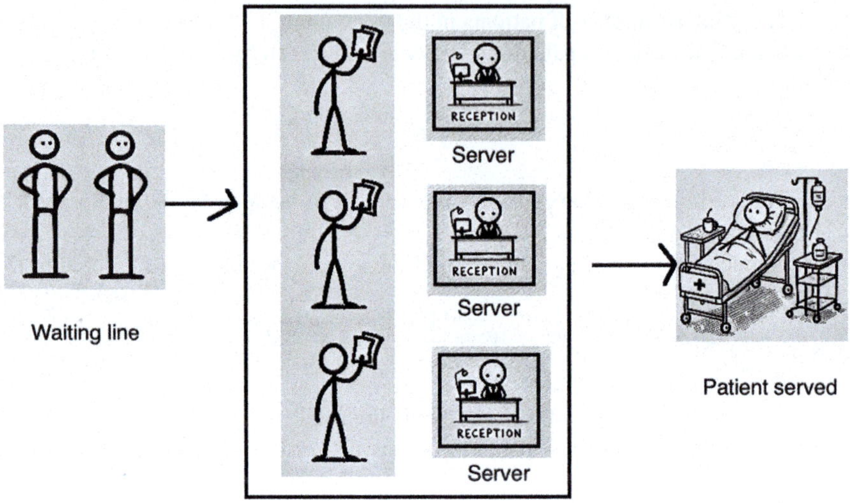


**Fig. 4.6** $M/M/k/FCFS$ model

- The queue discipline is first come, first served.

Let's consider we have an outpatient clinic where we have 3 doctors ($k$ servers). What are the key performance measures for this system? Let us answer the questions one by one. What do we know? We know that the arrival rate is 10 patients per hour, and the service rate is 5 patients per hour per doctor, hence $\lambda = 10$, $\mu = 5$.

Let's answer our questions.

- What are the chances (probability) of having no patients in the system?

$$P_0 = \frac{1}{\sum_{n=0}^{k-1} \frac{\left(\frac{\lambda}{\mu}\right)^n}{n!} + \frac{\left(\frac{\lambda}{\mu}\right)^k}{k!} \cdot \left(\frac{k\mu}{k\mu-\lambda}\right)} = 0.144$$

There are 14.4% chances that the outpatient clinic is empty.

- What is the average number of patients that are waiting?

$$L_q = \frac{\left(\frac{\lambda}{\mu}\right)^k \cdot \lambda \cdot \mu}{(k-1)! \cdot (k\mu - \lambda)^2} \cdot P_0 = 0.89$$

The average number of patients in the queue is 0.89 patients.

- What is the average number of patients in the whole system (patients that are waiting plus patients that are being treated in the meantime)?

$$L = L_q + \frac{\lambda}{\mu} = 2,89$$

The average number of patients in the system is 2.89.

- How much time does a patient have to wait to be seen by a doctor?

$$W_q = \frac{L_q}{\lambda} = 0.089$$

The average time that a patient has to wait to be seen by the doctor is 5.33 min.

- How much time does a patient spend in the system (waiting and service time)?

$$\begin{aligned} W &= \frac{1}{k!} \cdot \left(\frac{\lambda}{\mu}\right)^k \cdot \left(\frac{k\mu}{k\mu - \lambda}\right) \cdot P_0 \\ &= 0.29 \end{aligned}$$

The average time that a patient spends in the ER is 17.33 min.

- What are the chances of a patient that has just arrived to wait to be seen by a doctor, etc.?

$$P_{w>0} = \frac{(kp)^k}{k!\left(1 - \frac{\lambda}{\mu}\right)} \cdot P_0 = 0.444$$

There are 86.4% chances for a patient to wait to be seen by a doctor.

- What are the chances that at a certain moment to have $n$ patients in the system?

$$P_n = \left(\frac{k\lambda}{\mu}\right)^n \cdot \frac{1}{n!} \cdot P_0$$

The chances of having 0 patients in the system? 11.1%.

The chances of having 2 patients in the system? 22.2%.

The chances of having 4 patients in the system? 9.88%.

It is preferred to use multiple-channel systems, instead of multiple single-channel systems. Why? Because people jump from one waiting line onto the other, get frustrated if the other line moves faster, etc.

So far, we have discussed systems that assume that the interarrival time follows a Poisson distribution and that the service time follows an exponential distribution. What do we do if this is not the case in our situation? Luckily, Professor John Dutton Conant Little, from the Massachusetts Institute of Technology, developed the Little's law, that lets us compute the system's characteristic without having these presumptions mandatory. Little's law equations are:

*Little's law equations*

Let us create an example to illustrate Little's law equations. We have a clinic where patients arrive at an arrival rate of 12 patients per hour. On average, a patient spends in the clinic half an hour (0.5 h).

- We compute the average number of patients in the clinic as:

$$L = \lambda W = 12\frac{patients}{hour} \times 0.5\,\text{h} = 6$$

On average, in our clinic there are 6 patients waiting in line or being treated. Let's say we have two doctors in our clinic, $k = 2$, and that each doctor serves 4 patients per hour, $\mu = 4$. We can compute the fraction of time the doctors are busy treating patients as:

$$\rho = \frac{\lambda}{k\mu} = \frac{12}{2 \times 4} = 1.5$$

The utilization time is 1.5, which means that we have an overload in our system, the arrival rate exceeding the capacity of our doctors. This implies that the patients are arriving faster than they can be treated, which ultimately leads to an infinite queue. What is there to be done? Hire more doctors or redirect patients to another clinic. If we were to hire more doctors, let's say 2 more, then, the utilization time would be:

$$\rho = \frac{12}{4 \times 4} = 0.75$$

Meaning that each doctor is busy 75% of the time. We should adjust once more, since we are paying the doctors and 25% of the time they stay. Let's see what happens if we hire just 1 doctor:

$$\rho = \frac{12}{3 \times 4} = 1$$

Meaning that the doctors are busy 100% of their time. Perfect solution.

- The average number of patients that are waiting in queue:

$$L_q = \lambda W_q$$

We compute:

$$W_q = 0.5 - \frac{1}{4} = 0.25\,\text{h}$$

$$L_q = 12\frac{patients}{hour} \times 0.25\,\text{h} = 3patients$$

M/G/1 model

In this section, we will discuss the *M/G/1* model, in which the interarrival times follow a Poisson distribution, but the service time follows a general distribution $G(x)$. We will use the following notations:

- $\lambda$ represents the average arrival rate
- $\mu$ represents the average service rate
- $\frac{1}{\mu}$ represents the average service time
- $\sigma$ is the service time standard deviation.

We can compute the operating characteristics that help us answer our questions using the following formulas:

- The probability of having no patients in the system is:

$$P_0 = 1 - \frac{\lambda}{\mu}$$

- The average number of patients that are waiting is:

$$L_q = \frac{\lambda^2\sigma^2 + \left(\frac{\lambda}{\mu}\right)^2}{2 \cdot \left(1 - \frac{\lambda}{\mu}\right)}$$

- The average number of patients that are in the system (waiting in line plus being treated) is:

$$L = L_q + \frac{\lambda}{\mu}$$

- The average time that a patient has to wait to be treated is:

$$W_q = \frac{L_q}{\lambda}$$

- The average time that a patient spends in the system (waiting in line plus getting treated) is:

$$W = W_q + \frac{1}{\mu}$$

- The probability of a patient to wait to be treated when he/she/they arrive at the clinic is:

$$P_w = \frac{\lambda}{\mu}$$

Let us consider an example for a surgical department in a hospital. We have an operating room (O.R.) in which there are performed 2 surgical interventions per day. There are 8 working hours per day. The surgical interventions follow a Poisson distribution. The service times follow the Normal distribution with an average of 3.2 h and a standard deviation of 2 h. We are interested in finding out the following aspects: what is the number of patients that are waiting to be operated on, what is the average waiting time, and what is the operating time (%) in the O.R.?

We know that the average number of surgeries per hour is:

$$\lambda = \frac{2}{8} = 0.25,$$

while the mean service rate per hour is

$$\mu = \frac{1}{3.2} = 0.3125.$$

Therefore, the number of patients that are waiting for their surgeries will be computed as:

$$L_q = \frac{\lambda^2 \cdot \sigma^2 + \left(\frac{\lambda}{q}\right)^2}{2 \cdot \left(1 - \frac{\lambda}{\mu}\right)} = \frac{0.25^2 \cdot 2^2 + \left(\frac{0.25}{0.3125}\right)^2}{2 \cdot \left(1 - \frac{0.25}{0.3125}\right)} = 2.225$$

Having this result, we are able to compute the average waiting time, given by:

$$W_q = \frac{L_q}{\lambda} = \frac{2.225}{0.25} = 8.9\,\text{h}.$$

It should be noted that this time refers to the actual utilization of the O.R., which technically means that on average, a patient admitted in day $d$ will have the surgery on day $d + 1$.

Next, to compute the operating time (%) we need to compute the waiting probability, given by:

$$P_w = \frac{\lambda}{\mu} = \frac{0.25}{0.3125} = 0.80,$$

which means that 80% of the time spent in the O.R. is dedicated to performing surgeries.

If these numbers are not satisfactory, then the hospital manager should increase the mean service rate.

*M/G/k queuing model with Blocked Customers cleared (BCC)*

Another interesting queuing model is the M/G/k with BCC. In this type of model, we have multiple servers, but no queues are allowed. The interarrival times follow a Poisson distribution, while the service time is governed by the General distribution. If a patient arrives, and all the servers are occupied, the patient must leave, being erased technically from the waiting system. In general, there are two possibilities: either a patient leaves and then comes back later to check whether a server is unoccupied, or the patient leaves and never comes back. In this case, we need to ask ourselves how many servers we should have, so that we don't lose too many patients.

Let us imagine the following example. We have an online clinic, in which patients can call to ask for medical advice. Either they are instructed what to do at home, or are sent to the ER. We have a number of fixed telephone lines, so if all are occupied, the system does not allow callers to be put on hold. To compute what is the optimal number of telephone lines, we need to compute the probabilities of having exactly $j$ out of $k$ telephone lines occupied:

$$P_j = \frac{\left(\frac{\lambda}{\mu}\right)^j \cdot \frac{1}{j!}}{\sum_{i=0}^{k}\left(\frac{\lambda}{\mu}\right)^i \cdot \frac{1}{i!}}.$$

where $\lambda$ is the average arrival rate, $\mu$ is the average service rate for each server, and $k$ is the number of servers. To see what the percentage of blocked patient arrivals is, we need to compute what is the probability of having $k$ servers occupied, $P_k$.

We might be interested in computing the average number of patients equivalent to the average number of servers in use. For this, we compute:

$$L = \frac{\lambda}{\mu} \cdot (1 - P_k).$$

The other operating characteristics such as $L_q$ and $W_q$ cannot be computed, since now queue isn't allowed to be formed.

Let's take an example involving a medical help line. We presume that the mean number of calls is 12 calls per hour. Even if each call is different, in general 6 calls per hour can be handled. The medical help line has 3 internal telephone lines with a dispatcher. If a line is open, then the call is transferred directly to that dispatcher. If all lines are occupied, then the caller will hear the occupied tone and be disconnected. This is not an ideal situation. Let us see how we can model it, so that our medical help line takes 90% of the calls.

First, we will compute what is the probability of having all 3 internal telephone lines occupied:

$$P_3 = \frac{\left(\frac{12}{6}\right)^3 \cdot \frac{1}{3!}}{\left(\frac{12}{6}\right)^0 \cdot \frac{1}{0!} + \left(\frac{12}{6}\right)^1 \cdot \frac{1}{1!} + \left(\frac{12}{6}\right)^2 \cdot \frac{1}{2!} + \left(\frac{12}{6}\right)^3 \cdot \frac{1}{3!}} = 0.2105,$$

**Table 4.1** The probability of $j$ occupied lines out of 4

| Number of occupied lines | Probability |
|---|---|
| 0 | 0.1429 |
| 1 | 0.2857 |
| 2 | 0.2857 |
| 3 | 0.1905 |
| 4 | 0.0952 |

meaning that 21% of patient callers are turned away.

Let's see what the percentage of turned away patients is, if we add another internal telephone line, and we hire a new dispatcher.

$$P_4 = \frac{\left(\frac{12}{6}\right)^4 \cdot \frac{1}{4!}}{\left(\frac{12}{6}\right)^0 \cdot \frac{1}{0!} + \left(\frac{12}{6}\right)^1 \cdot \frac{1}{1!} + \left(\frac{12}{6}\right)^2 \cdot \frac{1}{2!} + \left(\frac{12}{6}\right)^3 \cdot \frac{1}{3!} + \left(\frac{12}{6}\right)^4 \cdot \frac{1}{4!}} = 0.0952,$$

meaning that we would lose only 9.52% of the patient callers, leaving 90.48% resolved cases.

Hence, the management problem is resolved if we open another line and hire a new person. Let us compute the average number of calls, that is the utilization rate of each telephone line:

$$L = \frac{\lambda}{\mu} \cdot (1 - P_k) = \frac{12}{6} \cdot (1 - 0.0952) = 1.81.$$

In Table 4.1, we computed the probabilities of having $j$ occupied lines out of 4.

Obviously, before installing a new line and hiring a new person, we should perform an economic analysis to see whether the costs of making this addition are lower than losing those patients. We need to see how much it costs to lose a caller and multiply that sum with the lost calls.

The number of lost calls per day is:
$8 \cdot 12 \cdot 0.0952 = 9.1$.

*Queuing model with finite number of potential patients*

So far, we have modelled situations where the number of patients was unlimited (ER, patient callers, etc.). Let us discuss now the situations in which we have a limited number of patients, for instance admitting people to a cardiology ward, where we have a limited number of beds. To compute the operational characteristics, we need to verify whether the following assumptions are true or not:

- We have only one server.
- The population is finite.
- The interarrival times follow the Poisson distribution.
- The service time is governed by the exponential distribution.
- The queue discipline is FCFS.

*Finite population M/M/1 queueing model*

All the computations will be made using $N$ as the size of the population. The following formulas will be used:

- The probability of having no patients in the system:

$$P_0 = \frac{1}{\sum_{n=0}^{N} \frac{N!}{(N-n)!} \cdot \left(\frac{\lambda}{\mu}\right)^n}$$

- The average number of patients that are waiting in line:

$$L_q = N - \frac{\lambda + \mu}{\lambda} \cdot (1 - P_0)$$

- The average number of patients that are in the system (waiting in line and being treated):

$$L = L_q + (1 - P_0)$$

- The average time that a patient needs to wait in line:

$$W_q = \frac{L_q}{(N - L) \cdot \lambda}$$

- The average time a patient spends in the system:

$$W = W_q + \frac{1}{\mu}$$

- The probability of having $n$ patients in the system:

$$P_n = \frac{N!}{(N-n)!} \cdot \left(\frac{\lambda}{\mu}\right)^n \cdot P_0, n = 0, 1, 2, \ldots N$$

Let us imagine the following scenario. In the radiology department, we have 5 nurses that bring patients for a CT scan and use a single CT machine. The average interarrival time for the CT scan is 40 min, the arrivals follow a Poisson distribution, hence we have the arrival rate:

$$\lambda = \frac{1}{40} = 0.025$$

On average, a CT scan is done in 5 min, the service times follow an exponential distribution, and the mean service rate is:

$$\mu = \frac{1}{5} = 0.20.$$

Hence, we have the traffic intensity equal to:

$$\frac{\lambda}{\mu} = 0.125.$$

Let us compute the operational characteristics. The probability that there are no patients waiting to get a CT scan is:

$$P_5 = \frac{1}{\sum_{n=0}^{5} \frac{5!}{(5-n)!} \cdot (0.125)^5} = 0.4790.$$

- The average number of nurses that are waiting in line with a patient is:

$$L_q = N - \frac{\lambda + \mu}{\lambda} \cdot (1 - P_0) = 0.3110.$$

- The average number of nurses that are waiting in line with a patient that are in the system (waiting in line and in the CT machine):

$$L = L_q + (1 - P_0) = 0.8320$$

- The average time that a nurse who is waiting with a patient needs to wait in line:

$$W_q = \frac{L_q}{(N - L) \cdot \lambda} = 2.9846$$

- The average time a nurse who is waiting with a patient spends in the system:

$$W = W_q + \frac{1}{\mu} = 7.9846$$

Knowing that there are 8 working hours, we wish to know how much time a nurse spends waiting for the patient to get the CT scan, and how much time the nurse spends to wait for the CT machine to be unoccupied. Hence, we perform the following computations:

$$8 \cdot 60 \cdot \lambda \cdot W = 95.8 \text{ min/day using the CT}$$

$$8 \cdot 60 \cdot \lambda \cdot W_q = 35.8 \text{ min/day using in line}$$

We are ending the presentation of the classical queuing models that are used for inpatient and outpatient management. In this chapter, we have provided multiple queuing models with examples that can be adapted to real life situations. In the next chapter we shall continue discussing queuing theory models that involve swarm intelligence optimization, models which can be successfully used for optimizing stock and supplies.

## References

1. Erlang, A.K.: The theory of probabilities and telephone conversations. Nyt Tidsskrift for Matematik, **20**(B), 33–39 (1909)
2. Anderson, D.R., Sweeney, D.J., Williams, T.A.: An Introduction to Management Science. West Publishing Company (1994)
3. Osaki, S.: Applied Stochastic System Modeling, Springer-Verlag (1992)

# Chapter 5
# Optimizing Stocks and Supplies

**Abstract** Profit in the detriment of patients: this is the worst nightmare. On the other hand, money do not grow in trees, medical insurances have high prices, and most people cannot afford medical treatment. How can we optimize stocks and supplies, so that we can offer the best possible medical care, without hemorrhaging money? Is this possible with Artificial Intelligence?

The last chapter left us discussing queueing theory models for inpatient and outpatient management. There is more to queuing theory than managing the time a patient has to wait in line to get treated. We need to discuss the optimization of stock and supplies. When managing a healthcare facility, we need to pay attention to money also. If we order too many supplies or we have too many beds, we lose money, money that could have been redirected somewhere else. On the other, having too little supplies or not enough beds, it means that we cannot treat people, so we redirect them to other healthcare facilities.

Between 1960 and 1970, two management theories have been proposed, Feldstein's theory and Avery-Jones' theory. Feldstein's idea suggested that there is a link between bed shortage and bed usage, and that link is connected to behavioral theory, [1]. In his perception, the doctors are under pressure to admit patients, hence there is an overload of patients, and not enough beds. He suggested that doctors should change the discharge behavior and stop building up waiting lists. Avery-Jones predicted that as medicine advances, the length of stay (hence the number of occupied beds) decreases, [2]. To make this prediction he used regression to analyze the length of stay of patients in beds. Statistical analysis of the two methods' performances proved that for the last 60 years, Avery-Jones theory was better than Feldstein's. As times change, do the results of previously developed theories also change? Does Avery's theory still stand today? For optimization purposes should we focus more on the bed usage with all the stocks and supplies that accompanies it, or on the speed of treatment and recovery?

We shall start by presenting another approach that allows a more flexible strategy to improve hospital management, [3]. This type of model is a system under the



S. Belciug, *The Hospital Manager's Guide to Artificial Intelligence*, Intelligent Systems Reference Library 263, https://doi.org/10.1007/978-3-031-80314-7_5

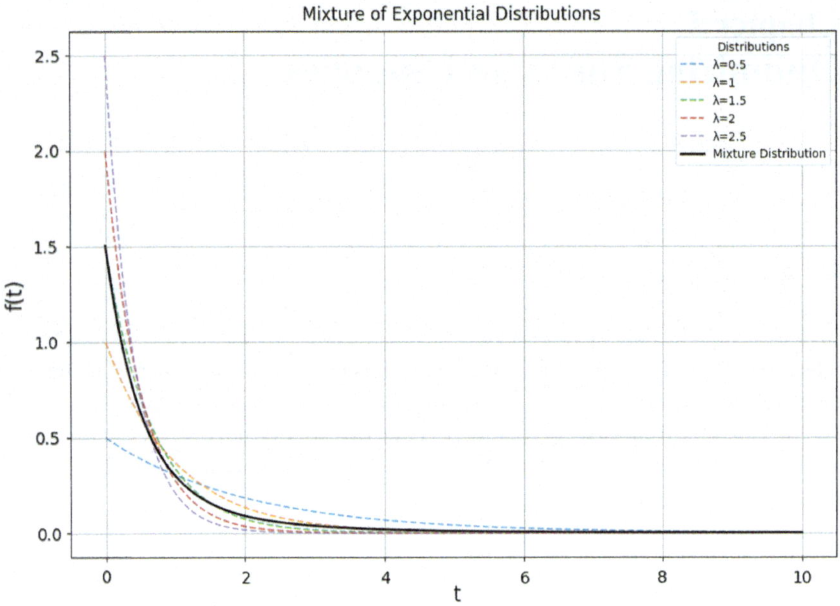


**Fig. 5.1** Mixture of exponential distributions

form *M/PH/c*. The interarrival times follow a Poisson distribution, but the service distribution is phase-type, [4], and the number of servers is *c*.

*Phase-type distributions*

Phase-type distribution has been developed since 1900s, starting with Erlang, [5], and was further discussed and improved by Jensen, [5], Neuts, [6], and Asmussen, [7], in 2003. To understand Phase-Type distribution you need to know Markov chains. If you've skipped Chap. 2, we strongly suggest you go back and read it, before continuing this section.

Let us suppose we have a Markov chain: the time distribution to absorption into a finite state (0) is a phase-type distribution (PH). If we keep things simple, we can think of PH distributions as *mixtures of exponential distributions* and *convolutions*. A *mixture distribution* is a merger between multiple distributions. The resulting distribution computes the probability of obtaining a predefined value, when randomly selecting one of the parent distributions and then obtaining a value from that parent distribution. Figure 5.1 presents a mixture of 5 exponential distributions, with different rate parameters, $\lambda$. The dotted lines represent the individual distributions, while the solid black line is the resulting mixture distribution. In this case, each component distribution weights equally in the mixture distribution.

In Fig. 5.2, we present a 3D plot of a mixture of exponential distributions. We can see how the components combine into a single one.

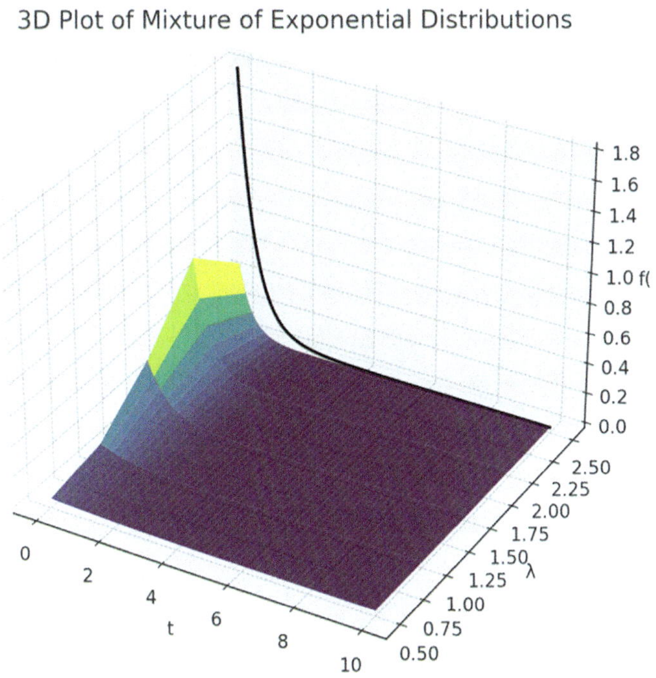


**Fig. 5.2** 3D plot of a mixture of exponential distributions

A *convolution* combines two probability distributions into one. The resulting probability distribution computes the probability of obtaining a certain value when adding the two parent distributions together. Figure 5.3 plots such a convolution. You can see the black curve that is the combination of the effect of the two distributions, resulting into a distribution that has a distinct peak and tail behavior. This is the impact of a convolution.

These distributions can be transformed into a Markov chain with a set of states and a transition matrix, [8].

When performing a Markov chain analysis, we assume that the waiting times follow an exponential distribution. So, the PH distribution and exponential distribution are interdepended. We denote with $PH(\alpha, S)$ the time distribution $X$ until the process reaches the absorbing state. The $X'$ CDF is given by:

$$F_X(t) = 1 - \alpha \cdot e^{\mathbf{T}t} \cdot 1$$

where

- $e^{\mathbf{T}t}$ is the matrix exponential of $\mathbf{T}t$. $\mathbf{T}$ is a subgenerator matrix, that is a matrix that describes the transition rates between transient states.
- $\alpha$ is the initial distribution, a vector that represents the initial probabilities of starting in each transient state.

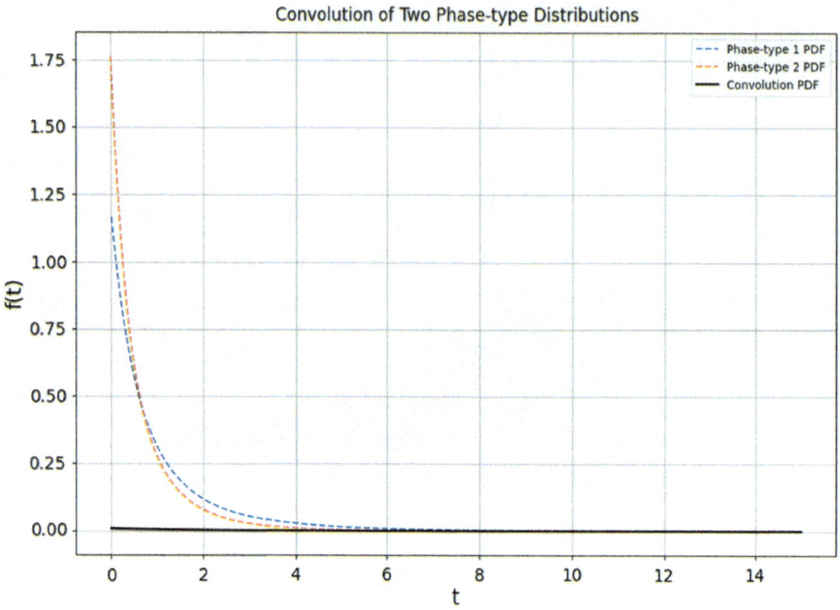


**Fig. 5.3** Convolution of two PH distributions

- $t$ is the absorption vector, a vector that represents the rates of absorption from each transient state to the absorbing state.
- $\alpha \cdot e^{\mathbf{T}t}$ is the probability distribution over the transient state at time $t$.
- 1 is a column vector of 1 s.

The $X$' PDF is:

$$f(x) = \alpha \cdot \mathbf{T} \cdot e^{\mathbf{T}t}\mathbf{1}.$$

Let us see if we can use the CDF and PDF of PH distribution to optimize the stocks and supplies in a hospital.

Can we model the patient's length of stay in the hospital as a Markov process? Yes, because it is a process that passes through different stages: admission, treatment, recovery, and finally the absorbing stage: discharge. The time a patient spends in each state is exponentially distributed, hence the whole process can be modelled as phase-type distribution.

Can we use the CDF and PDF in this case? Yes. The PDF will give us the probability density of a patient's length of stay at a moment in time $t$. We can use this to compute the rate at which patients are more likely to be discharged at different predefined moments in time. On the other hand, the CDF computes the cumulative probability of a patient to be discharged by the time $t$. Hence, we can forecast the number of patients that will be discharged in a time interval.

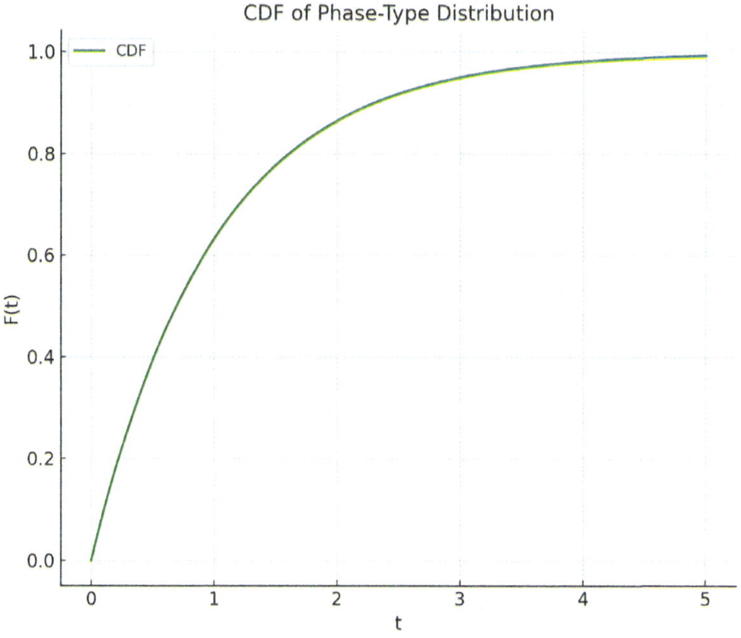


**Fig. 5.4** CDF for phase-type distribution

Therefore, we can use the PDF to forecast how many discharges will be made in the next period of time, and thus we can predict with a certain probability how many unoccupied beds we will have. If the discharge rate is high, then we can prepare for new admissions. For instance, if $F_X(3) = 0.8$, that is 80% of patients are expected to be discharged in the next 3 days, then we will know how many empty beds we will have after 3 days. We can combine this information, with intel regarding the demand. If we know that 100 patients will be admitted over the next 10 days, and 80% of the patients are expected to be discharged within this time, then we can estimate how many beds we still need. We can use this example to forecast certain treatments, what supplies are needed for them, etc. We can ensure that we do not run out of supplies, by anticipating when a phase is going to end, and replenish our stocks.

Figures 5.4 and 5.5 depict the CDF and PDF of a phase-type distribution.

Returning to our queueing model *M/PH/c*, we mathematically express the situation that all patients that find all the servers busy will be lost for the system as:

$$f(t) = \sum_{i=1}^{l} \alpha_i \rho_i e^{-\alpha_i t}$$

where

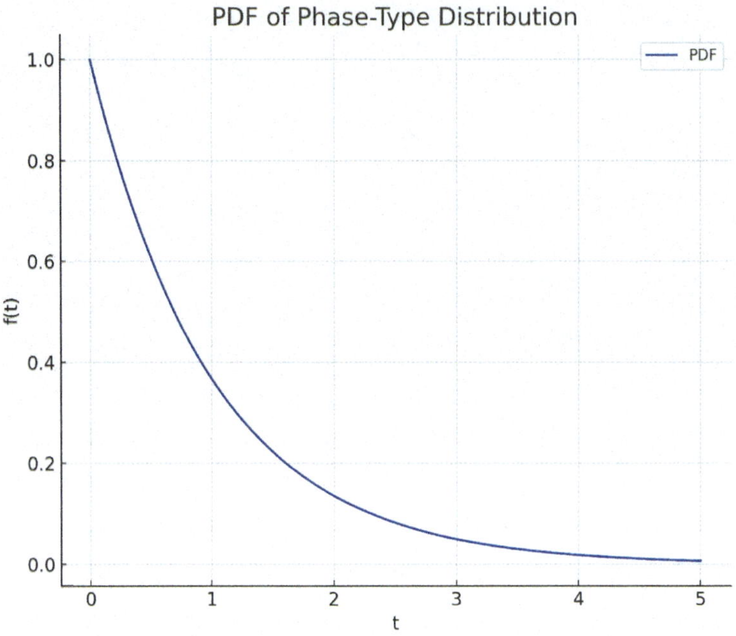


**Fig. 5.5** PDF for phase-type distribution

- $l$ represents the number of phases or compartments.
- $\alpha_i$ represent the mixing proportions.
- $\rho_i$ represent the transition rate. Please note that the sum of the transition rates must be 1, $\sum_{i=1}^{l} \rho_i = 1$.
- $\tau = \sum_{i=1}^{l} \frac{\rho_i}{\alpha_i}$ represents the corresponding mean.

Let us presume that given a time interval $t$, we have an average number of arrivals $\lambda \cdot t$. Thus, we can denote that the average number of arrivals during an average length of stay $\tau$ as $a = \lambda \cdot \tau$. The probability of having $j$ occupied servers is given by:

$$P_j = \frac{\frac{a^j}{j!}}{\sum_{k=0}^{c} \frac{a^k}{k!}},$$

while using Erlang's loss formula we can compute the probability of having all servers occupied as:

$$P_c = B(c, a) = \frac{\frac{a^c}{c!}}{\sum_{k=0}^{c} \frac{a^k}{k!}}.$$

The probability of having all servers occupied, and by servers we mean beds, treatment, etc., is equal to the fraction of patients that are lost by the system, $B(c, a)$. We are interested in three operational characteristics:

- $L = a \cdot [1 - B(c, a)]$ represents the average number of patients in the system. It is also known under the name *carried load*.
- $W = \tau \cdot [1 - B(c, a)]$ represents the average time spent in the system by a patient.
- $\rho = \frac{L}{c}$ represents the server occupancy. If our system in on steady state, then $\rho \leq 1$.

Can we provide the best of care for our patients at a minimum cost? Can we use our resources at the maximum level? The answer is yes, if we keep the potential loss of patients at a minimum with associate minimum costs. In this case, the waiting system will use an associated cost model that finds a balance between the service costs and loss patients, [9, 10].

For our systems to be successful, we will present different AI algorithm that help us optimize the resources. The optimization will determine the system's parameters $c, \lambda$, and $\tau$, so that we can compute the average time spent by the patients in the system, the average number of patients in the system, and a threshold for $B(c, a)$ that determines an appropriate proportion of refused patients tolerable by the system.

Moving further on, in the associated cost model, we can denote with $c$ the number of available resources (beds, drugs, personnel, etc.) and idle resources which can be used in case of emergency. Having this in mind, we need to add to more costs: a *penalty cost* for each $h$ unit/day/empty server and a *penalty cost* $\pi$ units/turned away patient. We can compute the cost/day using the following equation:

$$g(c, \lambda, \tau, h, \pi) = \pi \cdot \lambda \cdot B(c, \lambda \cdot t) + h \cdot \{c - \lambda \cdot t \cdot [1 - B(c, \lambda \cdot t)]\}.$$

Since $g(\cdot)$ represents the cost, in order to optimize the system, we need to minimize it. We need to find the optimum values for $c, \lambda, \tau, h$, and $\pi$. We know that the mean turned away patients per year is the ratio between the number of admissions and the number of allocated beds multiplied by 365, $T = 365 \cdot \frac{\lambda}{c}$, [11].

Using Markov chains, we can design the patient flow using compartmental models. Each phase determines a compartment. Initially, all patients are admitted to a certain first department, whether is acute care, or ICU, etc. A part of these patients recovers and are discharged, another part die, and another part is transferred to another department (for instance rehabilitation). From this new department the flow of patients resembles the one described so far: some are discharged, some die, and some get transferred elsewhere. For a better understanding we will present different models that have two or more compartments (Figs. 5.6).

In Fig. 5.6 we present a two-compartment model that shows the patient flow from the admission in the initial care unit, in this case the ICU, to a general ward. The arrows are indicating the direction of patient movement through the system. You can see that from the first compartment, the ICU, the patient can either go to the second department or die.

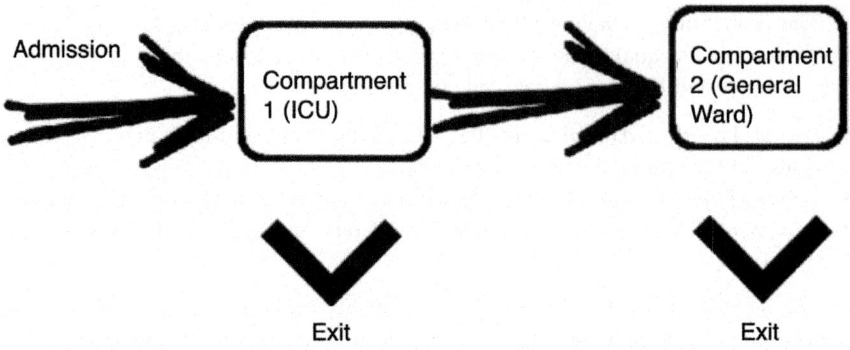


**Fig. 5.6** Two-compartment model for patient flow

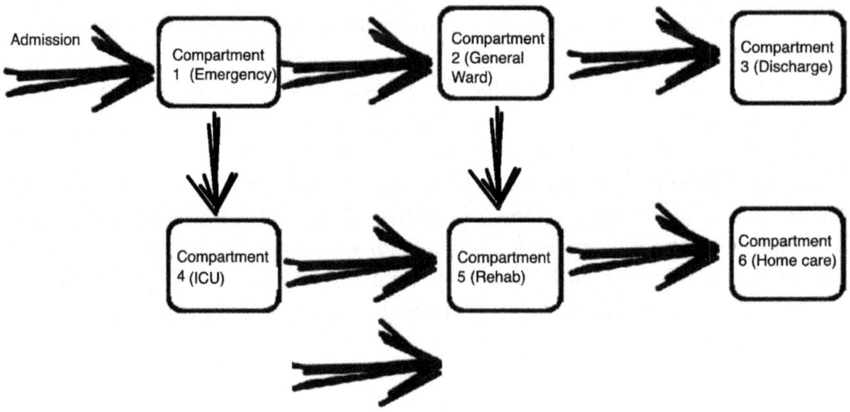


**Fig. 5.7** Complex patient flow compartment model

Figure 5.7 presents a more complex diagram of patient flow in a hospital. We have multiple departments that can be modeled using Markov chains. By modeling the patients flow we can understand and afterward optimize the system.

Next, we will present another example of a two-compartment model, [12]. We assume that a clinical department found in an equilibrium state has a constant admission rate $A$. Let us denote $N_i(s)$, $i = 1, 2$ the number of patients from ward $A$ and ward $B$, $s$ being the length of stay measured in days. We will consider the discharging rates from wards $A$ and $B$ as $r_1$ and $r_2$, with $r_2 < r_1$, and the transfer rate from ward $A$ toward $B$, $v_1$. Having these notations, we can describe mathematically the model as:

$$N_1(s + 1) = (1 - r_1 - v_1)N_1(s),$$

$$N_2(s + 1) = v_1 N_1(s) + (1 - r_2)N_2(s).$$

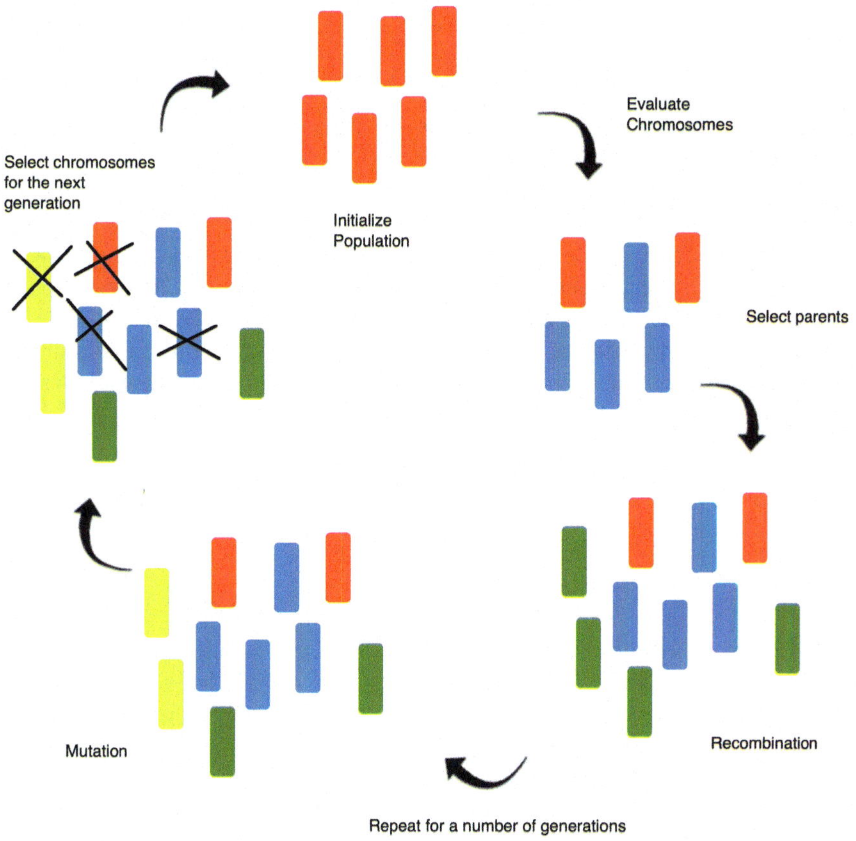


**Fig. 5.8** GAs diagram

| | | | | | | | | |
|---|---|---|---|---|---|---|---|---|
| 1 | 1 | 0 | 0 | 1 | 0 | 1 | 0 | Binary encoding |
| 8 | 2 | 1 | 5 | 6 | 3 | 4 | 7 | Integer encoding |
| 2.1 | 3.5 | 2 | 5.6 | 7.4 | 2.0 | 1.3 | 0.2 | Real encoding |

**Fig. 5.9** Chromosome representation: binary, integer, real-encoding

Depending on the parameters we need to optimize, the chromosome is *binary, integer*, or *real-encoded*. Figure 5.9 shows the three types of chromosome representations.

We select chromosomes to become parents, or form the next population based on a fitness function, fitness function that determines which are the individuals that are more likely to survive. It computes the capability of an individual to compete for survival purposes with the other individuals from the population. The value of the fitness function must be nonnegative. The overall fitness of the entire population that has $n$ chromosomes is given by:

$$F = \sum_{i=1}^{n} f(x_i).$$

There are several selection mechanisms for choosing the chromosomes that will produce offspring, [21, 22].

*The tournament selection*

When we use the tournament selection, we are making the individuals compete one on one to see which one gets selected to for recombination purposes. The first step is to select $k$ individuals, compute and compare their fitness values. The chromosome that has the best value (maximum or minimum depending on the optimization problem) gets selected. All chromosomes enter the fight having the same chances. A main disadvantage of this method is the fact that if we use it, the convergence speed might decrease, [23].

Each individual can be selected with a certain probability computed as:

$$P(i) = \begin{cases} \frac{C_{n-1}^{k-1}, \text{if} i \in [1, n-k-1]}{C_n^k} \\ 0, \text{if} i \in [n-k, n] \end{cases}.$$

We use the tournament selection when we have a large population. Having a large number of individuals in that population, it is hard to have knowledge regarding each competitor's knowledge and strength. Hence, we can select the best of them by making them compete one another (Fig. 5.10).

*Fitness proportional selection / Roulette-wheelselection/Monte Carlo roulette.*

In this type of selection, the fitness of each chromosome is set depending on its survival probability (Fig. 5.11). For this, we use the Monte Carlo roulette-wheel selection, in which a chromosome occupies an area on the roulette. The areas are proportional to the chromosomes' fitness values; hence they are not equal. The higher the fitness values, the higher the survival probability or the chance to get selected. We compute the probability of a chromosome to be selected as the ratio between its fitness value and the total fitness:

$$p_i = \frac{f(x_i)}{F}.$$

Because we have mentioned the Monte Carlo method, let us see how we can apply it:

1. At first, we need to identify with which type of distribution we are dealing with.
2. Secondly, we need to define the mean and standard deviation for each variable.
3. We artificially generate random data using the respective mean, standard deviation, and distribution.
4. The output of the process is estimated, and the results are analyzed.

The expected value of the random variable defines an integral, which can be approximated using the sample mean of the independent samples of the respective

**Fig. 5.10** The tournament selection illustration (generated using ChatGPT 4.0)

random variable. This approximation is possible due to the law of large numbers, which states that the mean result of an experiment obtained after a large number of trials is approximately equal to the expected value. As the number of experiments rises, the two values get closer.

***Ranking selection***

In this type of selection, the fitness of a chromosome determines its rank (Fig. 5.12). The differences between chromosomes are determined by ranks, and not by the fitness value. The probability of a chromosome to be selected is given by:

$$P(i) = \frac{rank(i)}{n \cdot (n-1)}.$$

*Exponential Rank selection*

In this type of selection, we compute the rank chromosomes according to their fitness and exponentially weight the obtained probabilities using a variable $c$, $0 < c < 1$. As the value of $c$ decreases towards 0, the exponentiality of the weight decreases

**Fig. 5.11** The Monte Carlo roulette selection (generated using ChatGPT 4.0)

**Fig. 5.12** Rank selection (generated using ChatGPT 4.0)

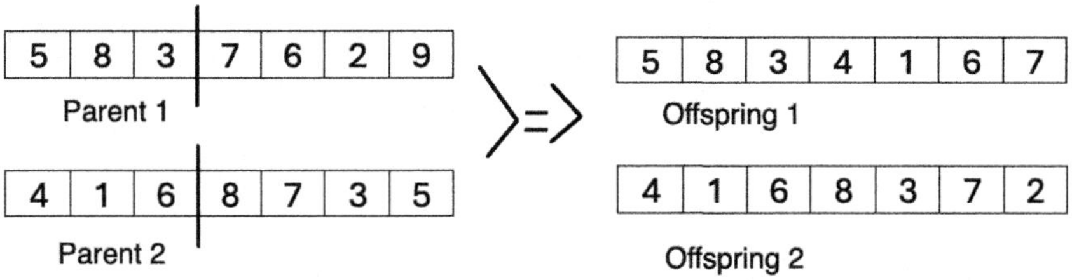


**Fig. 5.16** One-point crossover for integer representation

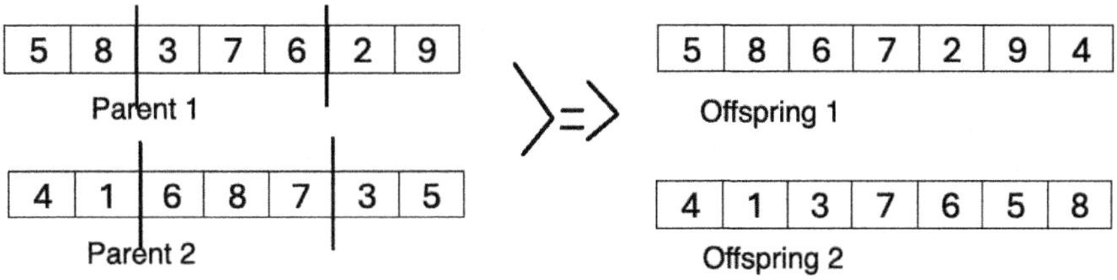


**Fig. 5.17** *N*-point crossover for integer representation

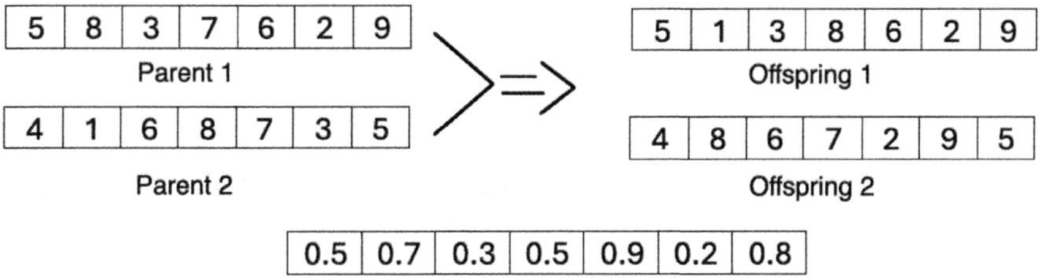


**Fig. 5.18** Uniform crossover for integer representation

- In the $n$-point crossover we keep the same principle as in the binary case, with the constraint that genes should not repeat in the offspring. Figure 5.17 presents the $n$-point crossover for integer representation.
- The uniform crossover for integer representation is the same as for the binary one. Nevertheless, we illustrate in Fig. 5.18 the process.

*Real-coded representation*

In the case our chromosomes are real-coded, we can use the following crossover mechanisms: simple arithmetic crossover, single arithmetic crossover, total arithmetic crossover, blend crossover, and Wright's heuristic crossover.

- In the *simple arithmetic crossover,* we randomly generate $k \in \{1, 2, \ldots, m - 1\}$, where $m$ is the number of genes in the chromosome. The first offspring is formed in the following manner: the first $k$ genes are copied from the first parent, whereas

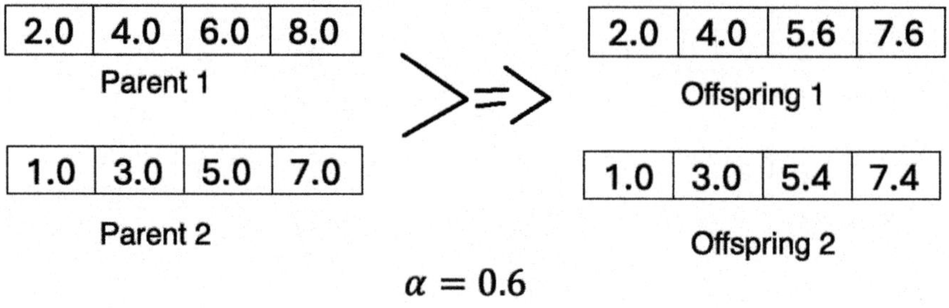


**Fig. 5.19** Simple arithmetic crossover

the remaining genes are computed using the following formula:

$$z_i^1 = \alpha \cdot x_i + (1 - \alpha) \cdot y_i, i = k + 1, \ldots, m.$$

The second offspring is formed using the first $k$ genes copied from the second parent, whereas the remaining genes are computed using the following formula:

$$z_i^2 = \alpha \cdot y + (1 - \alpha) \cdot x_i, i = k + 1, \ldots, m.$$

In Fig. 5.19 we present an example of the simple arithmetic crossover.

- In the *single arithmetic crossover*, we randomly generate $k \in \{1, 2, \ldots, m - 1\}$, where $m$ is the number of genes in the chromosome. To form the first offspring, we copy all the genes from the first parent, except gene $k$ which will computed as:

$$z_i^1 = \alpha \cdot x_k + (1 - \alpha) \cdot y_k.$$

The second offspring is formed by copied the genes from the second parent, except gene $k$, which will be computed as:

$$z_i^2 = \alpha \cdot y_k + (1 - \alpha) \cdot x_k.$$

In Fig. 5.20 we present an example that uses this type of crossover.

- In *the total arithmetic crossover,* we compute each single gene using the following two formulas:

$$z_i^1 = \alpha \cdot x_i + (1 - \alpha) \cdot y_i,$$

$$z_i^2 = \alpha \cdot y_i + (1 - \alpha) \cdot x_i.$$

In Fig. 5.21 we present an example on how to use this crossover method.

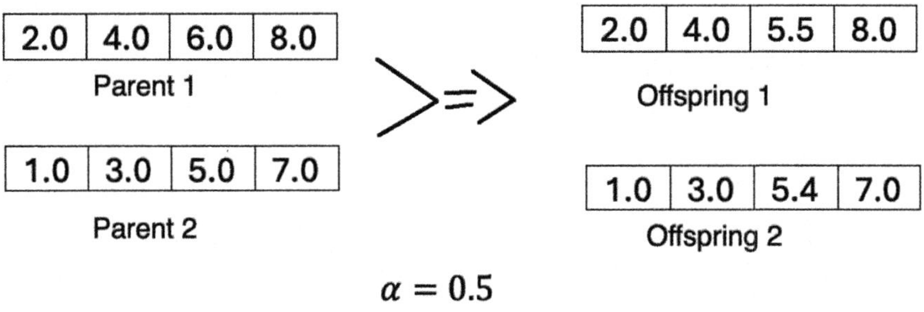


**Fig. 5.20** Single arithmetic crossover

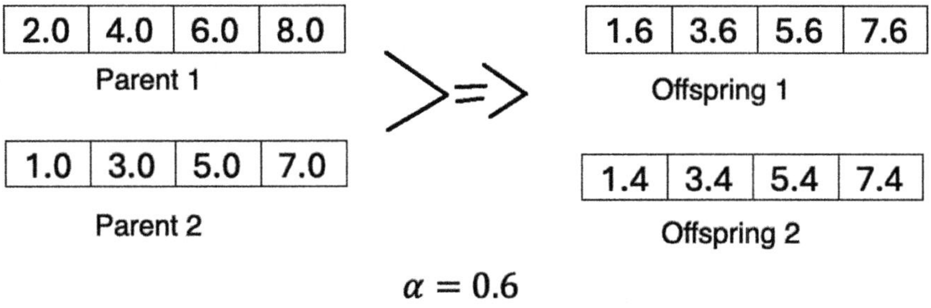


**Fig. 5.21** Total arithmetic crossover

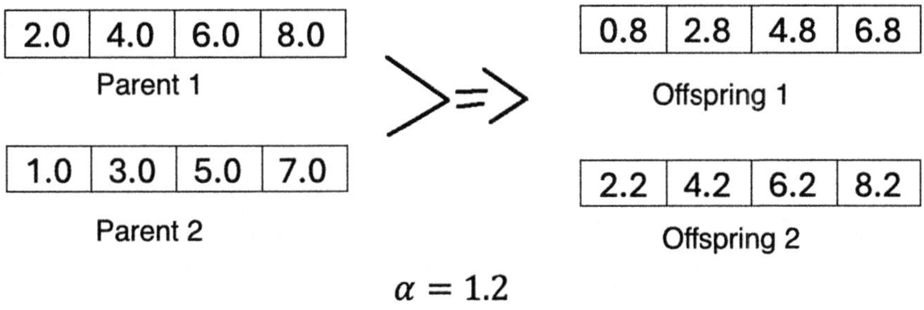


**Fig. 5.22** Blend crossover

- In the *blend crossover*, we create the offspring by blending the parents' genes. Let us suppose that the fitness score of parent $x$ is better than the fitness score of parent $y$, then the resulting offspring' genes will randomly generated from the interval $[x_i \pm \alpha \cdot (y_i - x_i), y_i \pm \alpha \cdot (y_i - x_i)$, if $x_i < y_i$.

  In Fig. 5.22 we present an example of how to compute the offspring.

- In the *Wright's heuristic crossover*, the offspring are formed using the parents' fitness scores. Let us presume that the fitness score of parent $x$ is higher than the fitness score of parent $y$, then the offspring is computed as, [27]:

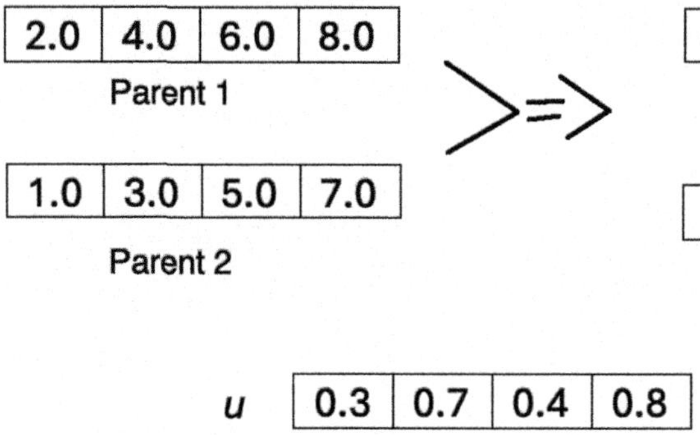


**Fig. 5.23** Wright's heuristic crossover

$$z_i^k = u \cdot (x_i - y_i) + x_i$$

In Fig. 5.23 we present an example of the Wright's heuristic crossover.

*Mutation*

We use the mutation operator to diversify our chromosomes, and to improve their performance. In this way, the evolution process is speeded up. The mutation can affect the entire chromosome or just parts of it. Depending on the representation we have multiple mutation mechanisms.

If our chromosomes have a binary representation, then we use the *bitwise mutation*. To see whether we mutate a gene or not, we randomly generate a number between 0 and 1. If that number is lower than the mutation probability, $p_m \in [0, 1]$, then we mutate that gene by changing its value from 0 to 1, or from 1 to 0, depending on the case. If the number's value is higher than the value of the mutation probability, then we do not mutate that gene, leaving the original value, 0 or 1.

If our chromosomes have an integer representation, we can choose between the *random setting mutation* and the *creep mutation*. The random setting mutation resembles the bitwise mutation: we generate a number between 0 and 1, and if that number is lower than the mutation probability, we replace the value of that gene with another value generated from the gene's value domain. If the number is lower than the mutation probability, then we leave it unchanged. The creep mutation uses two other generated values: a random value from [0, 1], that if it is below 0.5, we subtract, and if it greater than 0.5 then we add the other generated value. Please keep in mind that when subtracting or adding we must verify whether the obtain value is still inside the domain value.

If our chromosomes have a real representation, then we can use one of the two following methods: the *uniform mutation* and the *normally distributed mutation*. The uniform mutation changes the value of the gene with another one uniformly generated from the value domain, while the normally distributed mutation is similar to the integer creep mutation, the only difference being that the subtracting/added value is randomly generated using the Normal distribution with mean 0, and random standard deviation.

The genetic algorithm stops using a *stopping condition* that can be one of the following:

- Reaching a predetermined fitness value.
- Reaching a flat improvement of the fitness value.
- Reaching a predetermined number of generations.
- Reaching a limited diversity of the population.

More on how genetic algorithms can be used in optimizing stocks and supplies can be read in [3, 12].

A crucial part of using genetic algorithms for optimization problems is represented by the choice of parameters. We can either tune the parameters, by experimenting with different values and statistically validating the obtained results, or by changing the parameters dynamically during computer runs. Either way, we still need to use statistical tests to see the differences in performances. More on statistical tests in the next chapter.

***Artificial Immune Systems***

Another swarm optimization technique is artificial immune systems. This type of algorithm is inspired by the way the immune system works. There are three types of artificial immune systems: the clonal selection, the negative selection, and immune network. We are going to focus on the clonal selection algorithm, which has been developed by Burnet in 1957, [28].

Having as starting point the theory behind genetic algorithms, it is easier to understand the concepts in this new algorithm. We start by randomly generating a population of immune cells (antibodies) that go through a chancing process in response to a certain antigen. Each antibody represents a potential solution to our optimization problem. Each antibody goes through a selection mechanism that uses its affinity towards the antigen to see whether it is good or not. The best antibodies are selected and undergo a cloning and mutating process to improve their adaptive affinity toward the antigen.

Cloning an antigen means duplicating it a number of times. The number of clones depends on the antigen's affinity, the higher the affinity the higher the number of clones. After the clones are creating, they are hypermutated, inverse proportional to their affinity. The initial antibodies and the hypermutated clones go through another selection mechanism to form the next generation of antibodies. The low-affinity antibodies are replaced with new randomly created antibodies.

The process ends when the termination criterion is reached. One can use one of the stopping criterions from genetic algorithms.

More on how artificial immune systems can be used in optimizing stocks and supplies can be read in [29].

**Fig. 5.24** Transverse navigation system (image generated with ChatGPT 4.0)

***Moth flame optimization algorithm***

Another swarm intelligence approach that can be used in optimization problems is the moth flame optimizer. This algorithm was developed by Mijalili in 2015, [30]. It is inspired by the way a moth moves in the dark. This movement is known under the name of transverse navigation system. Moving in the dark, the moth needs to maintain a constant angle with a source of light (i.e. moon, candle, light bulb). The transverse navigation system can be modelled just like presented in Fig. 5.24.

Being a swarm intelligence algorithm, we are dealing with a population of moths, each moth representing a potential candidate solution. We mathematically represent a moth as:

$$X = \begin{bmatrix} X_1 \\ X_2 \\ . \\ . \\ X_N \end{bmatrix} = \begin{bmatrix} x_{11} & x_{12} & \ldots x_{1n} \\ \ldots & \ldots & \ldots\ldots \\ x_{N1} & x_{N2} & \ldots x_{N,n} \end{bmatrix},$$

where $N$ represents the number of moths in the population, and $n$ is the problem's dimension.

The flames are presented under a mathematical form as:

$$FM = \begin{bmatrix} FM_1 \\ FM_2 \\ . \\ . \\ FM_N \end{bmatrix} = \begin{bmatrix} Fm_{11} & Fm_{12} & \ldots Fm_{1n} \\ \ldots & \ldots & \ldots\ldots \\ Fm_{N1} & Fm_{N2} & \ldots Fm_{N,n} \end{bmatrix}.$$

The size of the flame matrix is equal to the size of the moth matrix.

To evaluate the moth and the flames, we are using associated fitness, expressed by:

$$f(X) = \begin{bmatrix} f(X_1) \\ f(X_2) \\ . \\ . \\ f(X_N) \end{bmatrix}$$

$$f(FM) = \begin{bmatrix} f(FM_1) \\ f(FM_2) \\ . \\ . \\ f(FM_N) \end{bmatrix}$$

In order for a moth to optimize its fitness, it needs to fly through the flame in a spiral movement. The spiral movement is modelled using the following logarithmic spiral function:

$$X_i^{K+1} = \delta_i \cdot e^{bt} \cdot \cos(2\pi t) + FM_i(k), \ i \leq N,$$

where $\delta_i = \left|X_i^K - FM_i\right|$ is the distance between the moth at position $X_i$ from its corresponding flame $FM_i$, whereas $b$ and $t$ are randomly generated from the [− 1, 1] interval. Using this formula, we are able to model the spiral movement shown in Fig. 5.25.

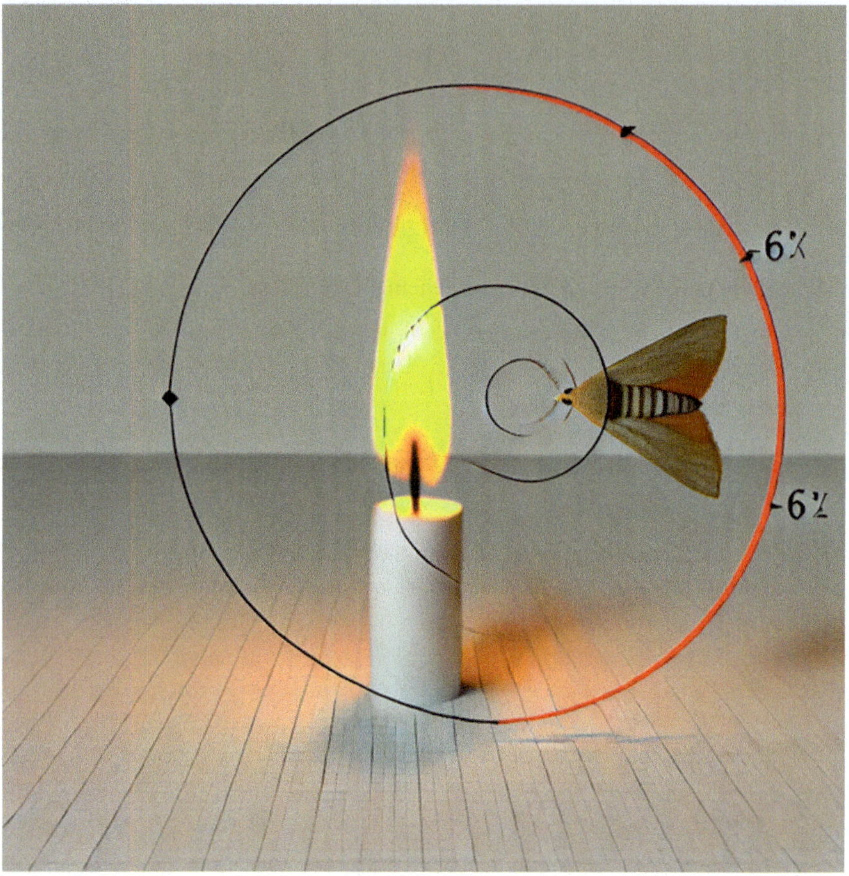

**Fig. 5.25** Logarithmic spiral movement of the moth toward the flame (image generated with ChatGPT 4.0)

Using the formula for the spiral movement, we compute the next position of the moth with respect to the flame. Now, let us discuss a little bit the parameters: $t$ controls how far or close the moth gets to the flame. If $t$ approaches 1, then the moth is closer to the flame, otherwise, the moth moves away from the flame. We store the best solutions obtained up until a moment in *FM* matrix. $t$ will be randomly generated form the interval $[r, 1]$, where $r$ is linearly increased from $-1$ to 0. We use $r$, the convergence constant, to ensure exploitation. To avoid getting trapped in a local optimum, the moth will update its position using only one flame. At each step, we sort the flames using their fitness values. We keep only the best flames and delete the rest. The following formula is used to reduce the number of flames:

$$no_{FM} = round\left(no_{FM_{lastiter}} - Current\_iter\frac{no_{FM_{lastiter}} - 1}{maxIter}\right).$$

Once the algorithm converges, the best solution (best flame plus moth) represents the final solution to our optimization problem.

More on how to use the moth flame optimizer to optimize stocks and supplies can be found in [31].

In this chapter we have presented different strategies on how to optimize the hospital's stocks and supplies using swarm intelligence. We shall move onto the next chapter where we are going to learn about Deep Learning and how to use it to improve the Pathology and Radiology department.

## References

1. Feldstein, M.S.: Effects on differences in hospital bed scarcity on type of use. BMJ 561–565 (1964)
2. Avery-Jones, F.: Length of stay in hospital. Lancet 321–332 (1964)
3. Belciug, S., Gorunescu, F.: Improving hospital bed occupancy and resource utilization through queuing modeling and evolutionary computation. J. Biomed. Inform. **53**, 261–269 (2015). https://doi.org/10.1016/j.jbi.2014.11.010
4. Faddy, M.: Examples of fitting structured phase-type distributions. Appl. Stoch. Models Data Anal. **10**, 247–2555 (1994)
5. Jensen, A.: A distribution model applicable to economics. Munksgaard, Copenhagen (1953)
6. Neuts, M.F.: Matrix-Geometric Solutions in Stochastic Models, John Hopkins Series in the Mathematical Sciences, vol. 2, .Johns Hopkins University Press, Baltimore (1981)
7. Asmussen, S.: Applied Probability and Queues. Springer Verlag, New York (2003)
8. Bean, N., Nielsen, B.: Decay rates of discrete phase-type distributions with infinitely-many phases. In: Matrix-Analytic Methods Theory and Applications. Proceedings of 4th International Conference, Adelaide, Australia (2002)
9. Gorunescu, F., McClean, S.I., Millard, P.H.: A queuing model for bed-occupancy management and planning of hospitals. J. Oper. Res. Soc. **53**(1), 19–24 (2002)
10. Stevenson, W.: Operation management, 10th edn. McGraw-Hill, Higher Education (2009)
11. Millard, P.H., Rae, B., Busby, W.: Why nosokinetics? Measuring and modeling the process of care. In: McClean, S., et al. (eds.) Intelligent Patient Management, SCI, vol. 189, pp. 3–23. Springer (2009)
12. Belciug, S., Gorunescu, F.: A hybrid genetic algorithm queuing multi-compartment model for optimizing inpatient bed occupancy and associated cost. Artif. Intell. Med. **68**, 59–69 (2016). https://doi.org/10.1016/j.artmed.2016.03.001
13. Stevenson, W.: Operations management, 10th ed. McGraw-Hill, Higher Education (2009).
14. Holland, J.H.: Adaptation in Natural and Artificial Systems: An Introductory Analysis with Applications in Biology, Control, and Artificial Intelligence. University of Michigan Press, Ann Arbor (1975)
15. Darwin, C.: On the Origin of Species by Means of Natural Selection, or Preservation of Favored Races in the Struggle of Life. John Murray, London (1859)
16. Radcliffe, N.J.: Genetic Neural Network on MIMD Computers (Unpublished D Phil Thesis). University of Edinburg, Edinburg (1990)
17. Watson, R.A., Pollack, J.B.: Recombination without respect: schema combination and disruption in genetic algorithm crossover. In: Whitley, D. Golberg, E.C.P,, Spector, L., Parmee, I., Beyer, H.G. (eds.) Proceedings of the 2000 Genetic and Evolutionary Computation Conference, Morgan Kaufmann, San Mateo, CA, pp. 112–119 (2000)
18. Schaffer, J.D., Whitley, D.L., Eshelman, L.J.: Combinations of genetic algorithms and neural networks: a survey of the state-of-the-art. In: Whitley, D.L., Schaffer, J.D. (eds.) Proceedings

of the International Workshop on Combinations of Genetic Algorithms and Neural Networks, pp. 1–37. IEEE Computer Socitey, Los Alamitos (1992)
19. Hancock, P.J.B.: Genetic algorithm and permutations problems: a comparison of recombination operators for neural net structure specification. In: Whitley, D.L., Schaffer, J.D. (eds.) Proceedings of the International Workshop on Combinations of Genetic Algorithms and Neural Networks, pp. 108–122. IEEE Computer Society, Los Alamitos (1992)
20. Whitley, D.L., Starkweather, T., Bogart, C.: Genetic algorithm and neural networks: optimizing connection and connectivity. Parallel Comput. **14**(3), 347–361 (1990)
21. Blickle, T., Thiele, K.: A comparison of selection schemes used in genetic algorithms (TIK-Report) **11** (1995)
22. Jebari, K., Madiagi, M.: Selection methods for genetic algorithm. Int. J. Emerg. Sci. **3**(4), 333–334 (2013)
23. Razali, N.M., Geraghty, J.: Genetic algorithm performance with different selection strategies in solving TSP. In: Proceedings of the World Congress on Engineering, II, UK (2011)
24. Eiben, A.E., Smith, J.E.: Introduction to Evolutionary Computing. Springer, Heidelberg (2003)
25. Eiben, A.E., Smith, J.E.: Multiparent recombination in evolutionary computing. In: Gosh, A., Tsutsui, S. (eds.) Advances in Evolutionary Computations: Theory and Applications, pp. 175–192. Springer, Heidelberg (2003)
26. Watson, R.A., Pollack, J.B.: Recombination without respect: schema combination and disruption in genetic algorithm crossover. In: Whitley, D., Golbder, E.C.-P., Spector, K., Parmee, I., Beyer, H.-G. (eds.) Proceedings of the 2000 Genetic and Evolutionary Computation Conference, pp. 112–119. Morgan Kaufamann, San Mateo, CA (1963)
27. Wright, A.: Genetic algorithms for real parameter optimization. In: Rawlins, G.J.E. (ed.) Foundations of Genetic Algorithms, pp. 205–218. Morgan Kaufmann, San Mateo, CA (1991)
28. Burnet, F.M.: A modification of Jerne's theory of antibody production using the concept of clonal selection. Aust. J. Sci. **20**(3), 67–59 (1957)
29. Belciug, S., Bejinariu, S.I., Costin, H.: An artificial immune system approach for a multi-compartment queuing model for improving medical resources and inpatient bed occupancy in pandemics. Adv. Electr. Comput. Eng. **20**(3), 23–30 (2020)
30. Mirjalili, S.: Moth-flame optimization algorithm: a novel nature-inspired heuristic paradigm. Knowl. Based Syst. **89**, 228–249 (2015)
31. Belciug, S., Ispas, F.: Smart optimization techniques for patient management. Kes InMed (2024)

# Chapter 6
# Pathology and Radiology Departments Improved by Artificial Intelligence

**Abstract** In this chapter, we are exploring how Artificial Intelligence algorithms can improve the Pathology and Radiology departments, by setting a fast and accurate diagnosis. False positives or false negatives can appear do to time pressure or fatigue. Artificial Intelligence lends a helping hand in this area through different methods. Both the Pathology and Radiology departments are always overcrowded. Waiting lists are building up, people live with the fear of a deadly diagnosis, thinking that by prolonging the time they are waiting to get a CT or an MRI scan, or wait for the pathology results, they will lose their fighting chance. Time is of an essence here, and every second counts. Artificial Intelligence acts like a second pair of eyes, guiding the doctor and red flagging them if they spot an abnormality. Theory plus examples are presented in this chapter.

Whenever a patient enters the ER or gets admitted in a hospital undergoes different tests: blood, imaging, or biopsies. The days when the doctor diagnosed with a stethoscope and some blood work routine are long gone. Nowadays, all diagnoses are set using images, whether we are discussing X-rays, ultrasound scans (US), computer tomography (CT) scans, nuclear medicine, Positron emission tomography (PET) CT mammography, magnetic resonance imaging (MRI), or pathology slides. The amount of work that goes on in the pathology or radiology departments is overwhelming.

Radiology is not used only to diagnose people, but to treat them, or monitor how the treatment works. Let us look at some numbers. In England, between February 2022 and January 2023, there have been reported 43.4 million imaging tests. In January 2023, there 3.41 million imaging tests were performed, out of which 1.65 million were X-rays, 0.81 million were US scans, 0.52 million were CT scans, and 0. 31 million were MRIs. If we look at global data, we see that the World Health Organization estimated that 3.6 billion diagnostic examinations are performed yearly. As time passes these numbers will rise even more.

By being able to process fast large amounts of images, AI solutions are promising to enhance the diagnostic accuracy, save time, and improve the patients' outcome. According to the American College of Radiology, 30% of radiologists already use AI in their clinical practice, and 20% are planning to use AI tools in the near future.

 
S. Belciug, *The Hospital Manager's Guide to Artificial Intelligence*, Intelligent Systems Reference Library 263, https://doi.org/10.1007/978-3-031-80314-7_6

The AI in radiology market size forecast shows an increase from $55.7 millions in 2021to $ 517.8 millions in 2030.

Here are some advantages in using AI in radiology:

- AI can detect diseases at earlier stages with a high level of accuracy, thus preventing further complications, considerably improving patients' outcome, and reducing the treatment costs.
- AI can provide an automatic triage of scans by prioritizing them according to the severity of the disease, thus saving time and letting doctors handle urgent cases first.
- By being trained on huge amounts of data, the AI tools learn to detect abnormalities in scans more accurately than human doctors, thus increasing patients' chance of surviving.
- AI is able to set the optimal dosage of radiation, reducing the level of unnecessary radiation for the patients.
- AI can reduce radiation exposure by enhancing imaging scans, thus repeated imaging becomes unnecessary.
- AI can enhance the image quality of scans, providing better detection and diagnosis.
- AI improves patients' satisfaction by offering a fast and accurate diagnoses, and thus allowing them to access treatment more quickly, improving their chances of survival.
- AI can automatically produce standardized reports streamlining the workflow and saving time.

Figure 6.1 provides all the above-mentioned advantages of AI in radiology at a glance.

Regarding pathology, the mean number of pathologists per million population is 14. But as we have learned so far, the mean does not always provide the correct representation of the reality. While in the United States, there are 65 pathologists per million, in Africa there are less than 3 pathologists per million. Reports state that over two thirds of the total number of pathologists around the world are clustered in only just 10 countries. 95% of patients are going to have a pathologist involved in their care at some point in their life. One way to resolve the disparities and reduce the pathologist's workload is to use AI to process the slides.

A good manager should make use of the on the market available AI tools to optimize the doctor's workload and improve patients' outcomes.

Now that we know that AI tools are robust and reliable to process images, let us see how it is done. To understand the new technologies, we first need to go back to the basis and learn how one of the most powerful machine learning techniques was developed, and how it evolved over time in order to become a trustworthy solution in modern medicine.

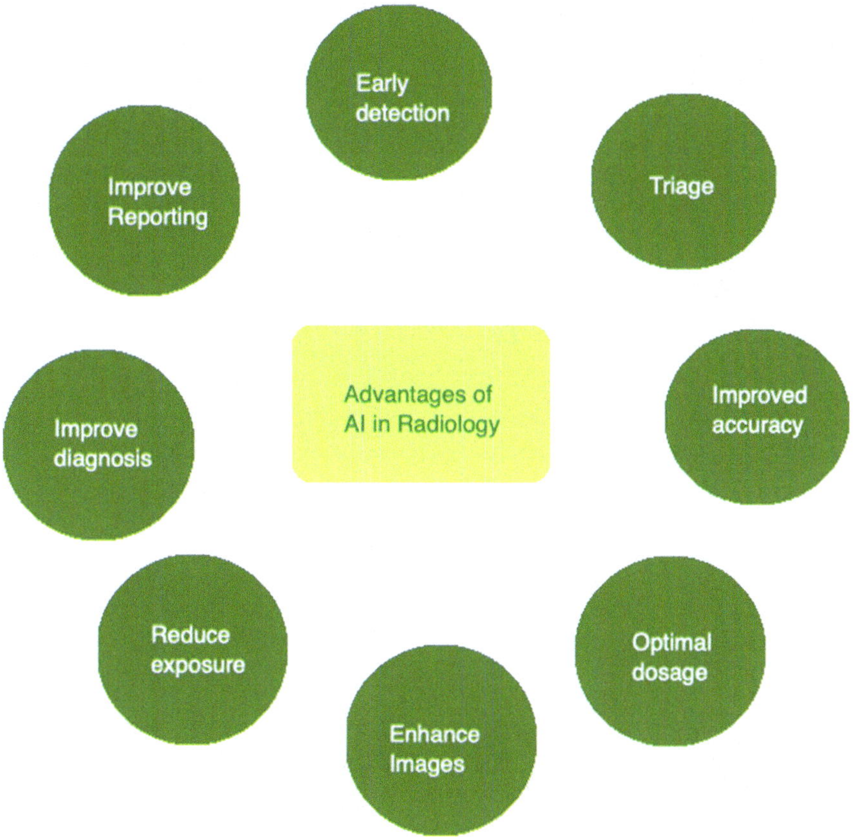


**Fig. 6.1** Advantages of AI in radiology

### *Artificial Neural Networks*

In Chap. 5, we saw that scientists inspired themselves in creating AI algorithms from nature. They studied how group intelligence can solve optimization problems. A very powerful AI method is the Artificial Neural Network (NN). As the name implies, artificial NNs are mimicking the way the human brain works, how we think, remember, create and innovate. In 1943, Walter Pitts and Warren McCulloch, developed the artificial neuron, [1]. The work was continued by Donald Hebb in his book "The Organization of behavior", [2], and by Rosenblatt, that designed the first artificial neuron, the *perceptron*. The perceptron uses a weighted sum and if its result surpasses a predefined threshold, then it outputs 1, otherwise it outputs 0. The weights are tuned by a learning process. Teaching children how to differentiate between animals resembles the way we teach the perceptron to tune its weights in order to minimize the error between its output and the ground truth.

Let's show how this process works in mathematical form. We shall denote the input as $x_i$, $i = 1, 2, \dots, n$. Each $x_i$ is weighted using the weight $w_i$, and summed up.

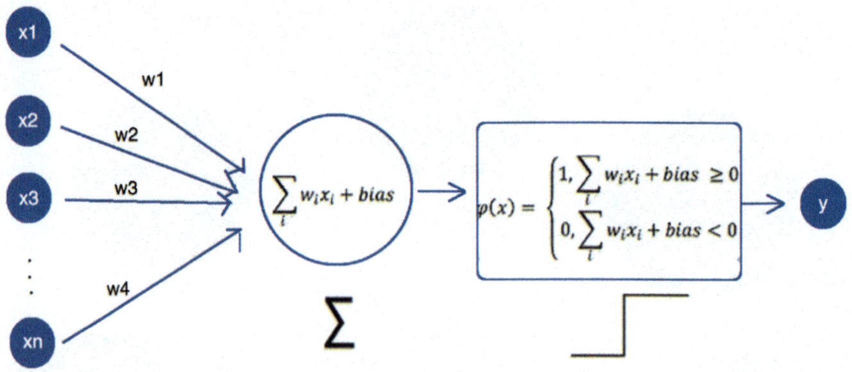


**Fig. 6.2** Rosenblatt's perceptron

We apply an activation function of our choice on the result and compare the output to the a priori predefined threshold. Figure 6.2 presents Rosenblatt's perceptron.

To compute a neuron's value, we use the following equation:

$$s_j = \sum_{i=1}^{p} w_{ij} \cdot x_i = \boldsymbol{w}_j \cdot \boldsymbol{x}',$$

where we denote the synaptic vector with $\boldsymbol{w}_j = (w_{j1}, w_{j2}, \ldots, w_{jp})$, the input with $\boldsymbol{x}_j = (x_{j1}, x_{j2}, \ldots, x_{jp})$, and the linear combiner with $s_j$. The summation result, $s_j$ is passed through the activation function that compares its result with the predefined threshold, to decide whether the neuron will fire or not. Mathematically speaking, the activation of a neuron is described by:

$$y_j = \varphi(b_j + s_j) = \begin{cases} h_j, \text{if } s_j + b_j \geq T_j \\ 0, \text{if } s_j + b_j < T_j \end{cases},$$

having denoted with $\varphi$ the activation function, and with $b_j$ the bias. $y_j$ is the computed output value. If the data is linear separable, then we obtain good results, otherwise the results are poor. Figures 6.3 and 6.4 show what linear and non-linear data mean.

In literature there are multiple activation functions that you can choose from. The most used activation functions are:

- The *Heaviside* or *threshold* activation function has the following formula and graph (Fig. 6.5):

$$\varphi(s) = \begin{cases} 0, s < 0 \\ 1, s \geq 0 \end{cases}.$$

This is the formula that was used by McCulloch and Pitts.

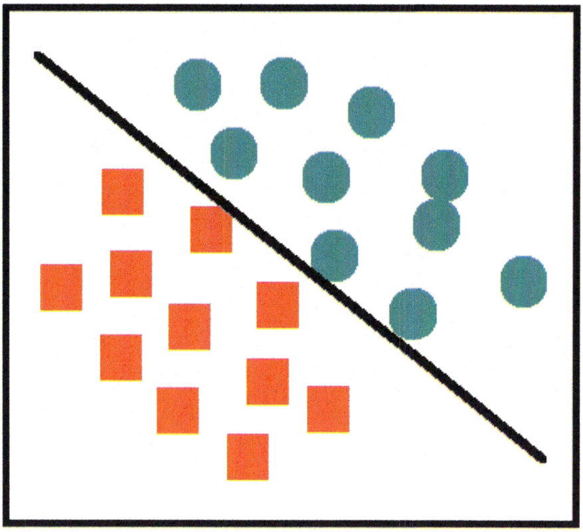

**Fig. 6.3** Linear separable data

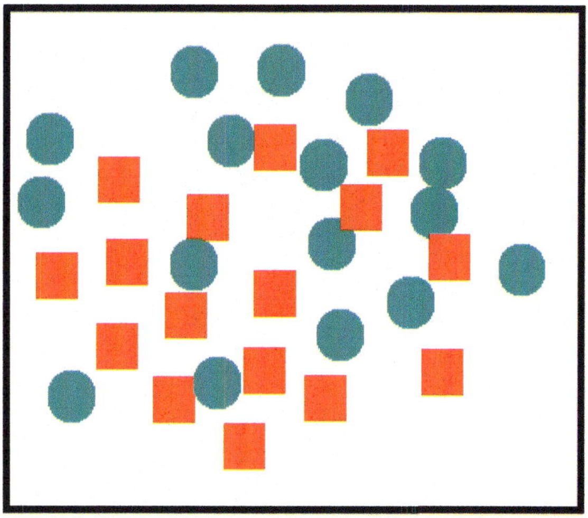

**Fig. 6.4** Non-linear separable data

If we replace 0 with -1 or use the *step activation function*, we obtain the *generalized Heaviside* (Fig. 6.6). Its formula is presented below together with its graph.

$$\varphi(s) = \begin{cases} -1, s < 0 \\ 1, s \geq 0 \end{cases},$$

$$\varphi(s) = \begin{cases} a, s < h \\ b, s \geq h \end{cases}.$$

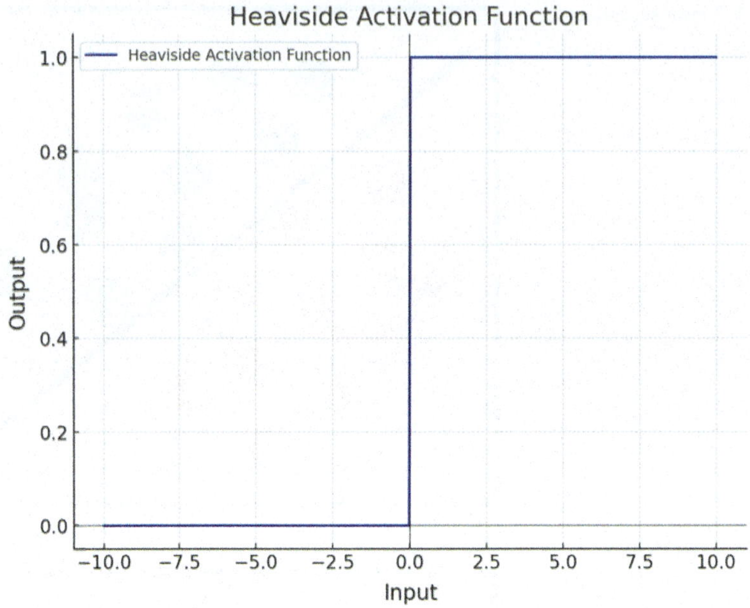


**Fig. 6.5** The Heaviside / threshold activation function graph

- The *piecewise-linear / ramp* activation function has the following formula and graph (Fig. 6.7):

$$\varphi(s) = \begin{cases} -a, s \leq -a \\ s, |s| < a \\ a, s \geq a \end{cases}.$$

- The *linear* activation function has the following formula and graph Fig. 6.8):

$$\varphi(s) = a \cdot s.$$

- The *Gaussian* activation function has the following formula and graph (Fig. 6.9):

$$\varphi(s) = \exp\left(-\frac{s^2}{a}\right).$$

- The *sigmoid* activation function has the following formula and graph (Fig. 6.10):

$$\varphi(s) = \frac{1}{1 + \exp(-a \cdot s)}.$$

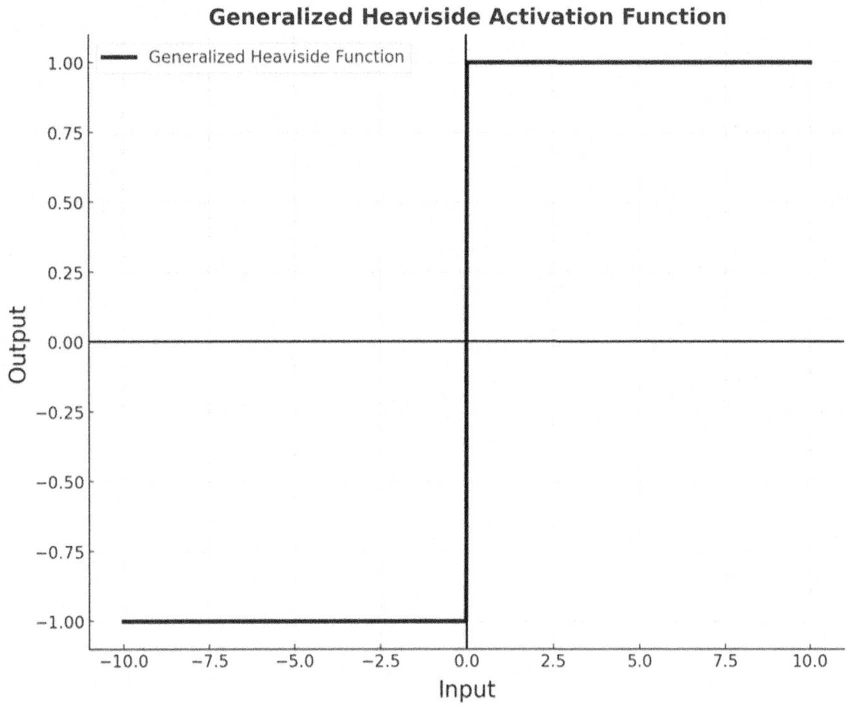


**Fig. 6.6** The generalized Heaviside function graph

We can see from the sigmoid's graph that the function takes a value from $\mathbb{R}$, and squashes it into a value that belongs to the interval (0, 1).

- The *hyperbolic tangent* activation has the following formula and graph (Fig. 6.11).

$$\varphi(s) = \tanh(s) = \frac{e^s - e^{-s}}{e^s + e^{-s}}.$$

If you look close enough at the hyperbolic tangent's graph you will see that it resembles to the sigmoid, the only difference being that it takes a real value and squashes it into a value that belongs to the interval (− 1, 1). This difference has an impact on your AI model's results, and you should know some tips and tricks so that you choose the correct activation function wisely.

The sigmoid is not zero centered, this implies that for inputs that are close to 0, the output will be around 0.5, whereas the tangent is zero centered, so any input close to 0 will produce a close to 0 output, which is a big plus for the next layers in a neural network to learn, because all the neurons' outputs are centered.

Another difference between the two functions is the gradient. The sigmoid's gradient has a flatter curve, which causes the vanishing gradient problem as we

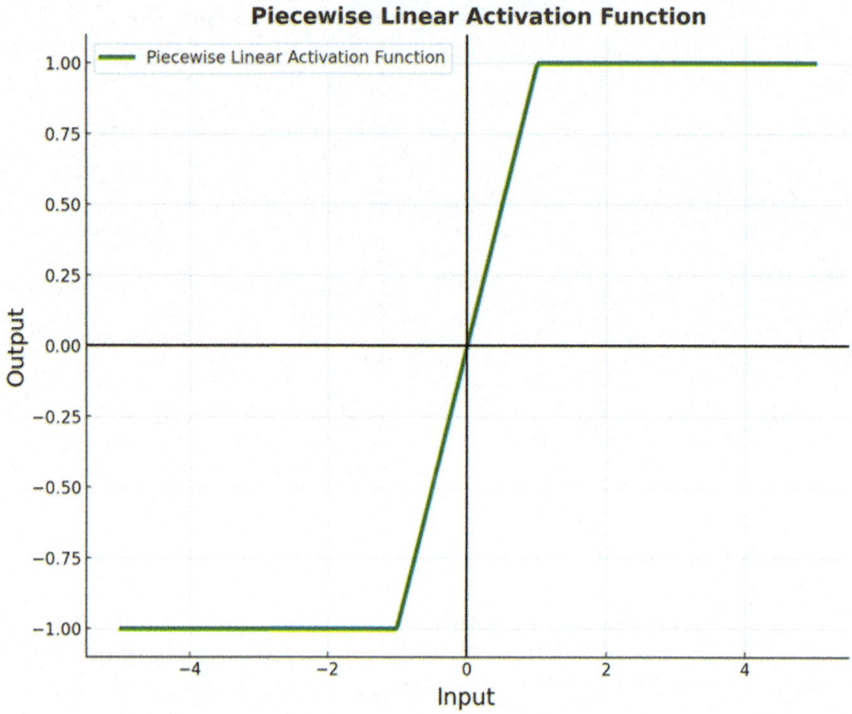


**Fig. 6.7** The piecewise activation function graph

move away from 0, which leads to a poor learning process for our NN. On the other hand, the tangent has a steeper gradient around zero, thus it has a stronger gradient, avoiding the vanishing gradient problem. More details regarding the gradient will be provided later on in this chapter.

You should use the sigmoid when you are dealing with binary classification tasks, or where the output needs to represent a probability, thus a value between 0 and 1. The tangent on the other hand is generally used in the activation of the hidden layers of the NN, if the data is normalized, because it is able to map the input to a range that is more effective for further processing. In Fig. 6.12 we have plotted the two functions together, so you should have a clearer picture regarding the differences.

- The *Rectified linear unit or ReLU* activation function has the following formula and graph (Fig. 6.13):

$$\varphi(s) = \max(0, s).$$

ReLU has become the most popular activation function in deep learning due to the following reasons:

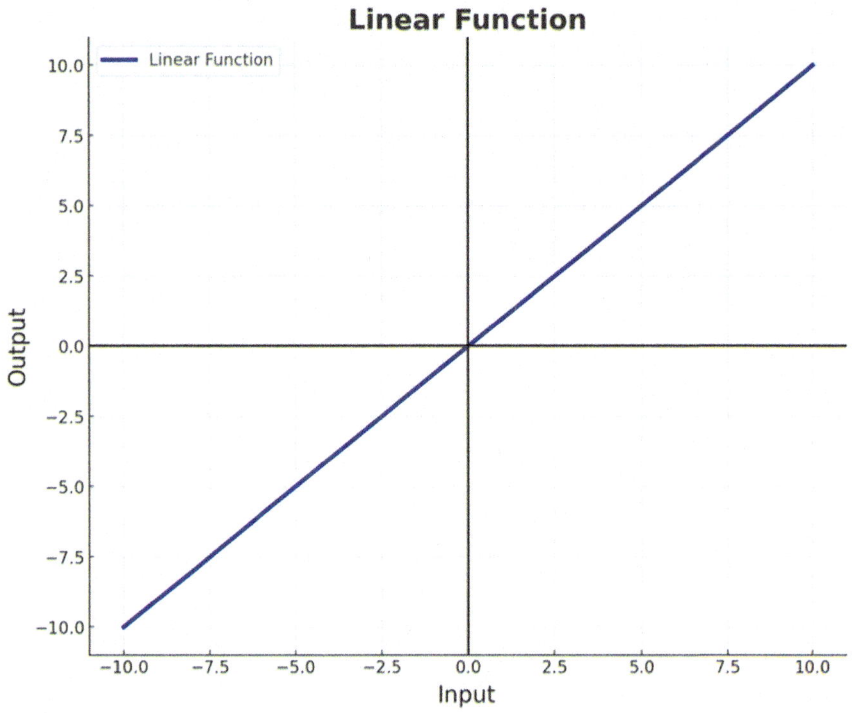


**Fig. 6.8** The linear activation function graph

- It is simple. ReLU has a simple mathematical formula: it returns 0 for any negative input, or the input itself for a positive one. This function is easier and faster than exponential functions such as the sigmoid or hyperbolic tangent.
- It is sparse. ReLU produces sparse activations, meaning that a significant number of neurons will have as output 0. ReLU sparsity leads to a more efficient model, because it reduces the amount of computations that need to be made, and also it reduces the probability of the network to overfit.
- It has a non-saturating nature. Recall that we have mentioned that both the sigmoid and tangent suffer from the vanishing gradient problem (i.e. the gradients are too small, hence the learning process is slowed down), ReLU does not. It does not saturate for positive inputs, letting the gradient remain large, and thus the NN to learn at a faster pace.
- It converges faster. ReLU converges faster during training than both the sigmoid and the tangent. This means that we need fewer epochs to train the network, hence it is more efficient in practice.
- It handles both positive and negative inputs. ReLU makes the NN learn with respect to positive inputs, thus making it learn important features more efficiently. By treating positive and negative inputs differently, the NN is able to capture complex patterns.

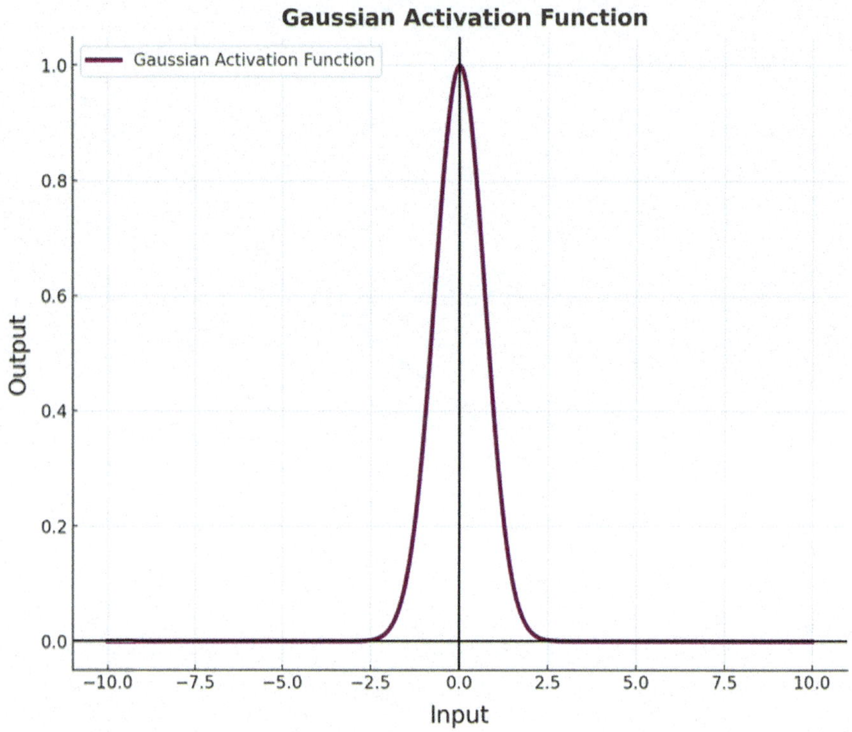


**Fig. 6.9** The Gaussian activation function graph

Obviously, it is not only milk and honey when it comes to ReLU. An advantage might be a disadvantage: to many neurons become inactive and only output 0. A potential solution to this problem is using *Leaky ReLU* or *Parametric ReLU* as activation functions. Both functions allow small non-zero gradients when dealing with negative input.

- The *Leaky ReLU* activation function has the following formula and graph:

$$\varphi(s) = \max(\alpha s, s),$$

  where $\alpha$ is a small positive number (i.e. 0.01). Leaky ReLU allows small non-zero output for negative input values (Fig. 6.14).
- The *Parametric ReLU (PReLU)* activation function has the following formula (Fig. 6.15):

$$\varphi(s) = \max(\alpha s, s),$$

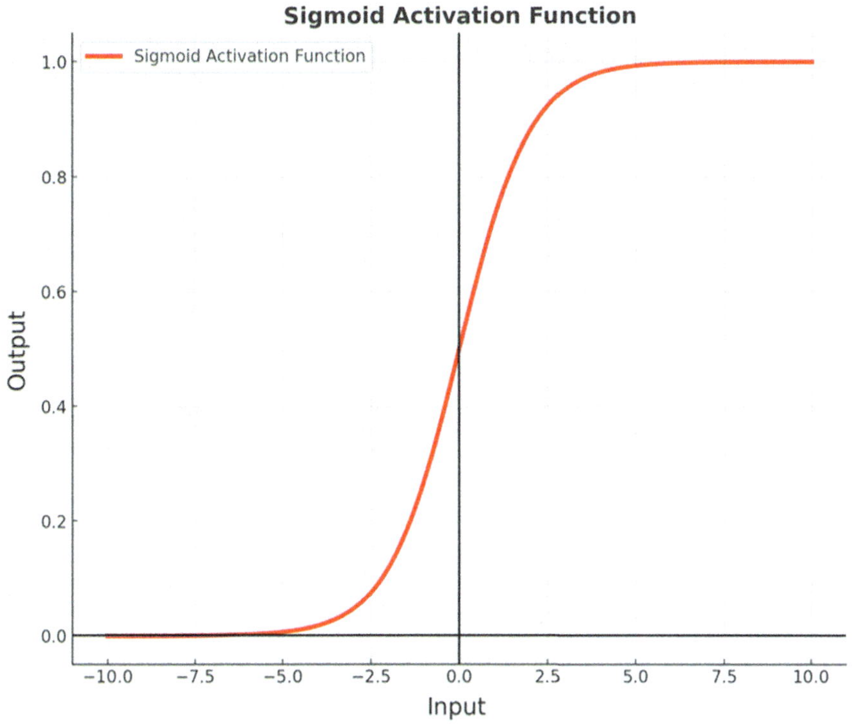


**Fig. 6.10** The sigmoid activation function graph

where $\alpha$ is a parameter learned during training. Unlike Leaky ReLU that has a fixed slope for negative inputs, the PReLU's slope is a parameter adjusted during training, allowing the NN to adjust its behavior taking into account data, which might lead to an improvement in performance. A potential drawback is represented by the increased complexity due to the additional parameter. If not handled properly it may lead to overfitting. Figure 6.14 presents the ReLU, Leaky ReLU, and PReLU graphs plotted together, for you to see better the differences between them.

As time passed, NNs have become more complex, by adding more neurons and more layers. The number of neurons, the number of layers, how they are structured, and how the signal propagates throughout the network represents the architecture of the NN. In what follows we will present different NN architectures.

In every single NN we have three types of layers: the *input* layer, the *output* layer, and the *hidden* layers between them. The signal can travel throughout the network from input to output—*forward propagation*. In this case, we are dealing with a *feedforward* NN. If our networks have loops, then we are dealing with a *recurrent* NN. Figures 6.16 and 6.17 present a feedforward NN architecture, and a recurrent NN architecture respectively.

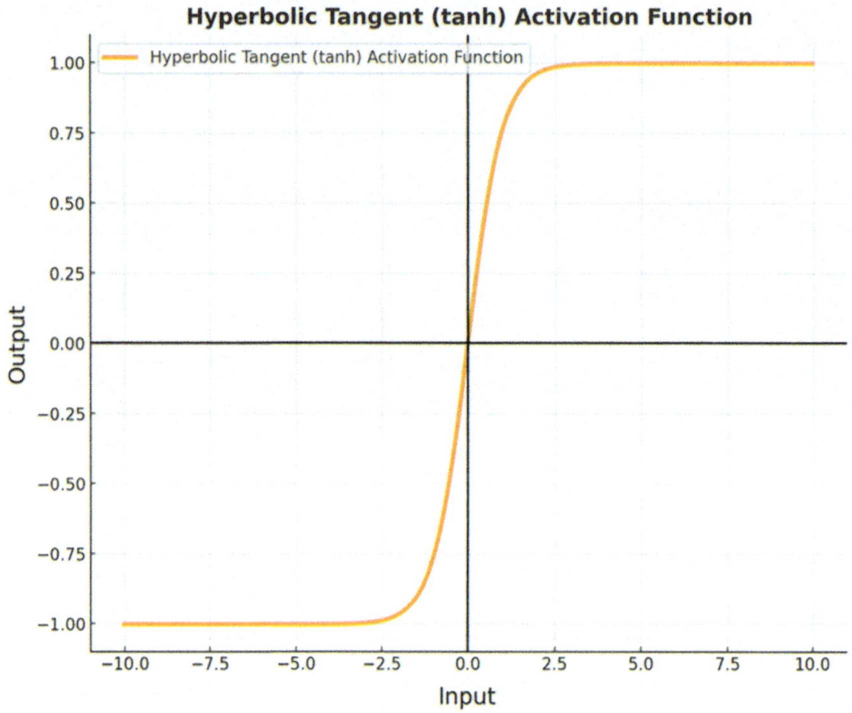


**Fig. 6.11** The hyperbolic tangent activation function graph

Depending on the way the neurons are organized in layers we can have the following architectures:

- The *single-layer feedforward NN* has only an input and an output layer. There are no hidden layers in this type of NN. It can solve linearly separable problems (Fig. 6.18).
- The *multi-layer NN* or the *multi-layer perceptron* (MLP) is the extension of the *single-layer NN*. It contains an input layer, an output layer, and multiple hidden layers in between, (Fig. 6.19). Each layer is fully connected to the next. The outputs of a layer are the inputs of the next. MLPs are able to solve non-linear problems, that is data that cannot be separated by a line. If your NN has enough hidden neurons and layers it can approximate any continuous function. Deep learning is a type of MLP, (Fig. 6.20).
- The *recurrent NN* (RNN) is a network that has loops in it. This type of architecture allows the network to have memory regarding its previous inputs. RNN are useful when we are dealing with data where the order of inputs is important (i.e. time series, speech recognition, natural language processing).

The activation functions between the input layer and the hidden layers produce real outputs. The activation function between the last hidden layer and output has

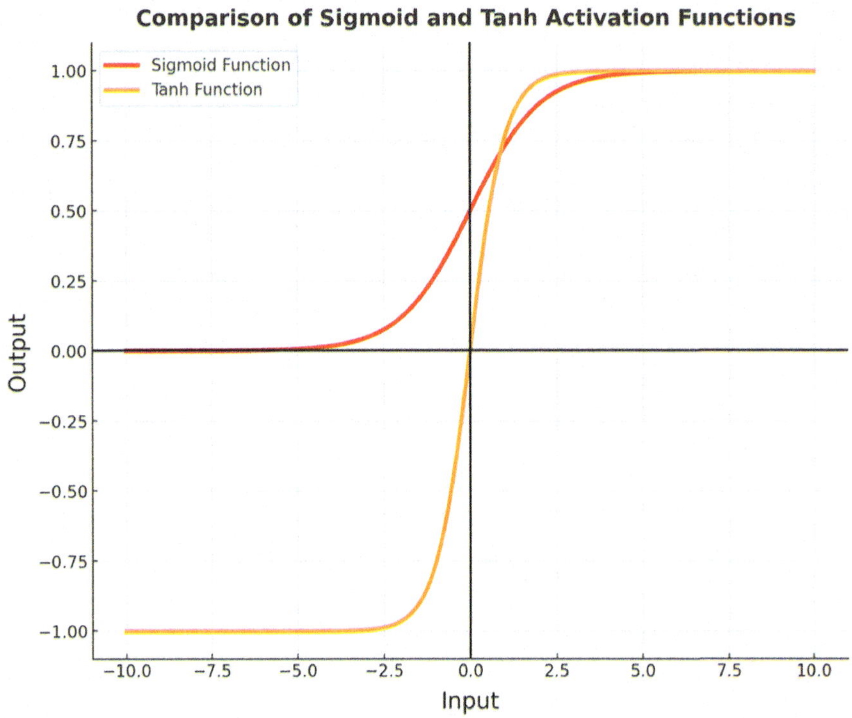


**Fig. 6.12** Comparison between the sigmoid and the hyperbolic tangent activation functions

to transform a continuous real value into a categorical one. If we have a two-class decision problem, then we can use the logistic regression which takes a real value and transforms it into a binary one. If we are dealing with a multiple-class decision problem (> 2 classes), we use the *softmax* activation function.

*Logistic regression*

Logistic regression is used to compute the probability of an output, having as input different explanatory variables. The output is transformed into a binary value through the use of $logit(p)$, where $p$ is the proportion of items with the same attribute. For instance, $p$ is the probability of a person to have cancer. Logit is computed using the following formula:

$$logit(p) = \ln\left(\frac{p}{1-p}\right) = b_0 + b_1 \cdot X_1 + b_2 \cdot X_2 + \cdots + b_m \cdot X_m.$$

Let us denote $a = b_0 + b_1 \cdot X_1 + b_2 \cdot X_2 + \cdots + b_m \cdot X_m$. The we compute $p$ as:

$$p = \frac{1}{1 + e^{-a}}.$$

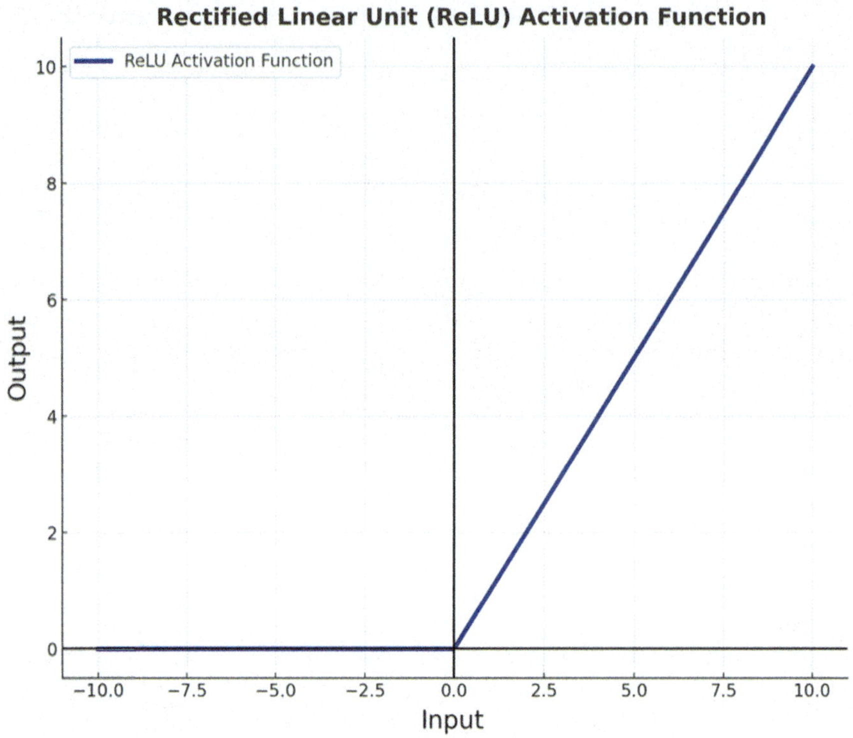


**Fig. 6.13** The ReLU activation function graph

*Softmax*

The softmax activation function takes a continuous real value and transforms it into probabilities. The sum of these probabilities equals 1. Probabilistically speaking, softmax computes the probability distribution of the hypothetical outcomes:

$$s_i(y) = \frac{e^{y_i}}{\sum_j e^{y_j}}.$$

To apply softmax, we need to *one-hot encode* the classes. The one-hot encoding process converts each output into an array of zeros and one. The length of the array is given by the number of classes. The value 1 is placed on the index that matches the number of the class. Let's give an example regarding staging cancer: class 0 – no cancer, 1 – cancer grade 1, 2—cancer grade 2, 3—cancer grade 3, and 4—cancer grade 4.

- Class 0: $y_0 \rightarrow [0, 0, 0, 0, 1]$;
- Class 1: $y_1 \rightarrow [0, 0, 0, 1, 0]$;
- Class 2: $y_2 \rightarrow [0, 0, 1, 0, 0]$;

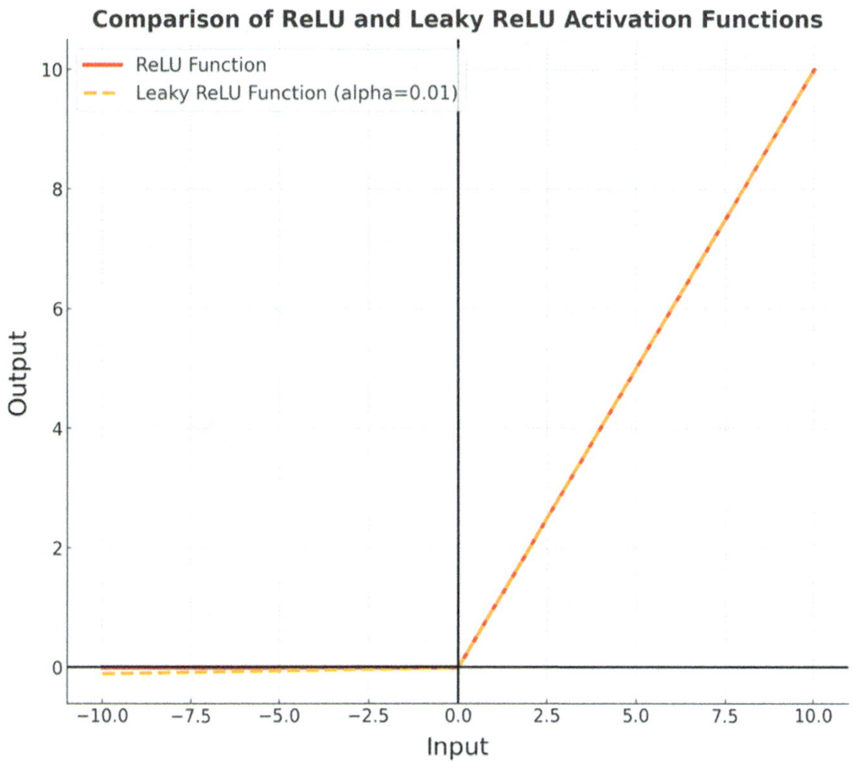


**Fig. 6.14** The ReLU versus Leaky ReLU activation functions graph

- Class 3: $y_3 \rightarrow [0, 1, 0, 0, 0]$;
- Class 4: $y_4 \rightarrow [1, 0, 0, 0, 0]$;

Another way to decide which class is the correct one is by applying the *winner-takes-all* rule and selecting the highest value.

Let's apply both softmax and the winner-takes-all on the same example, and see what happens. Let us presume that our network outputs the following raw scores (logits):

- Class 0: 0.8
- Class 1: 1.2
- Class 2: 0.5
- Class 3: 2
- Class 4: 1.5

According to the winner-takes-all rule the class that has the highest raw score is chosen as the output. Since class 3 has the highest score (2.0), the NN will predict class 3. All the other classes are ignored, and only class 3 is considered as the final prediction.

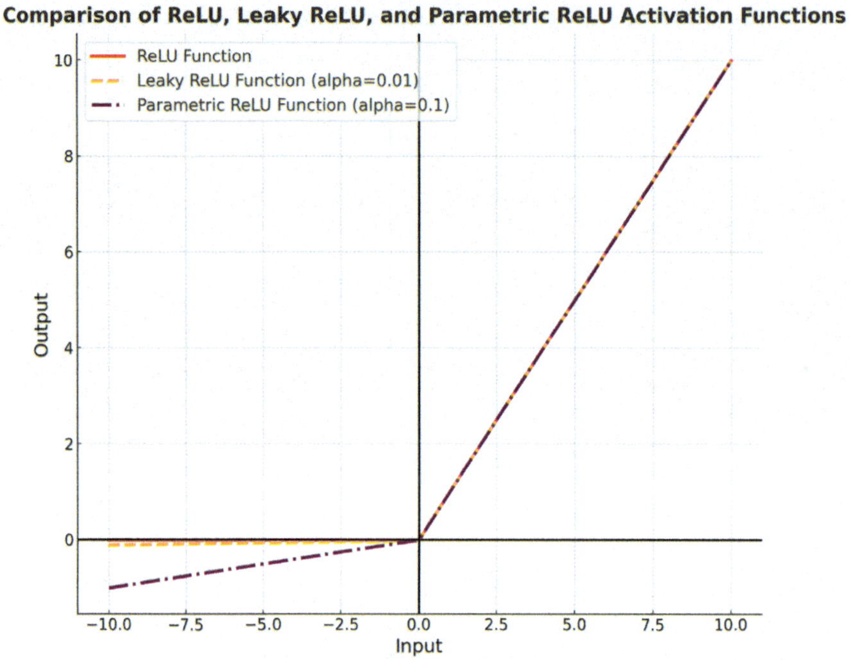


**Fig. 6.15** Comparison between ReLU, Leaky ReLU, and PReLU activation functions graph

Softmax will transform the raw scores into probabilities. This will give us insights about how confident the NN regarding each class is to be the correct prediction.

Let us compute the softmax values. First, we need to exponentiate the raw scores:

- Class 0: $e^{0.8} \approx 2.225$
- Class 1: $e^{1.2} \approx 3.320$
- Class 2: $e^{0.5} \approx 1.649$
- Class 3: $e^{2.0} \approx 7.389$
- Class 4: $e^{1.5} \approx 4.482$

Now, we need to sum up the exponentials:

$$2.225 + 3.320 + 1.649 + 7.389 + 4.482 \approx 19.065$$

The final step is for us to compute the softmax probabilities:

- Class 0: $\frac{2.225}{19.065} \approx 0.117(11.7\%)$
- Class 1: $\frac{3.320}{19.065} \approx 0.174(17.4\%)$
- Class 2: $\frac{1.649}{19.065} \approx 0.086(8.6\%)$
- Class 3: $\frac{7.389}{19.065} \approx 0.387(38.7\%)$
- Class 4: $\frac{4.482}{19.065} \approx 0.235(23.5\%)$

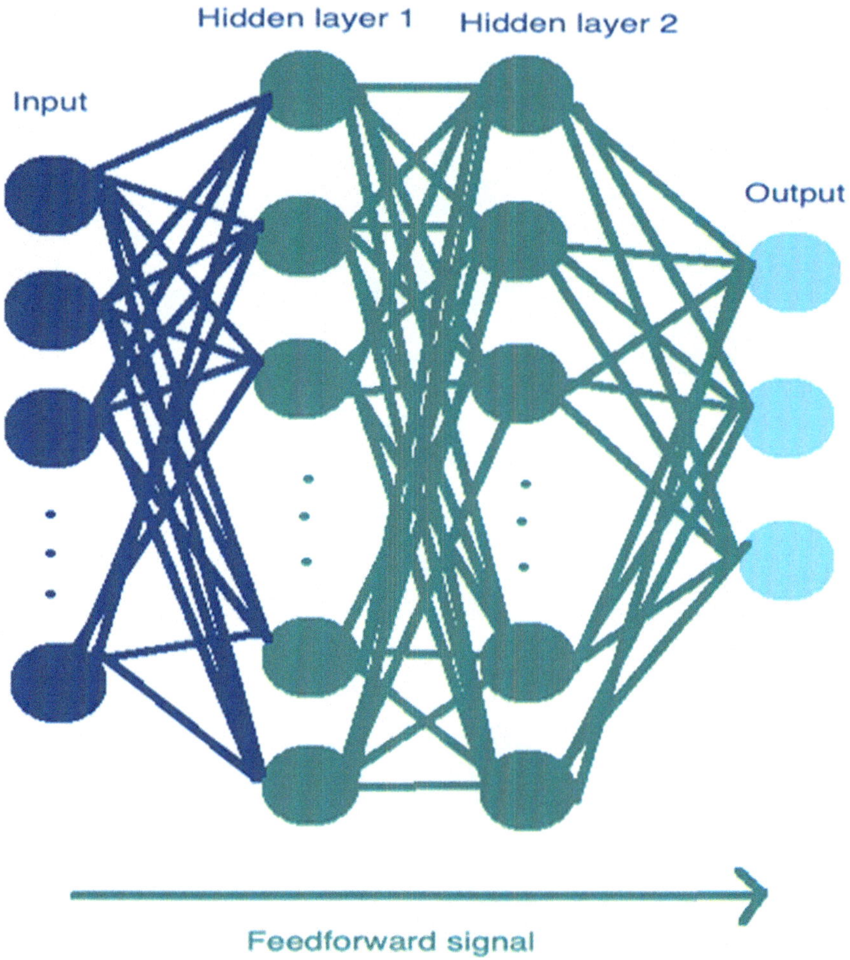


**Fig. 6.16** Feedforward neural network

The softmax probabilities show that while class 3 is the most likely to be the correct one (38.7%), the NN assigned a high probability to class 4 also (23.5%). The other classes have smaller probabilities. Hence, the conclusion is the following: both methods gave the same result, class 3, but softmax offered us more insights into the model's confidence in its predictions, which might come in handy in our decision-making process.

In Fig. 6.21, we have plotted the softmax function. We can see that there is a smooth transition across the input values, which reflects the softmax function's behavior in shifting the probabilities. This smoothness is very useful when we have close input values. You can see from the plot that it has a sharp increase (from the exponential)

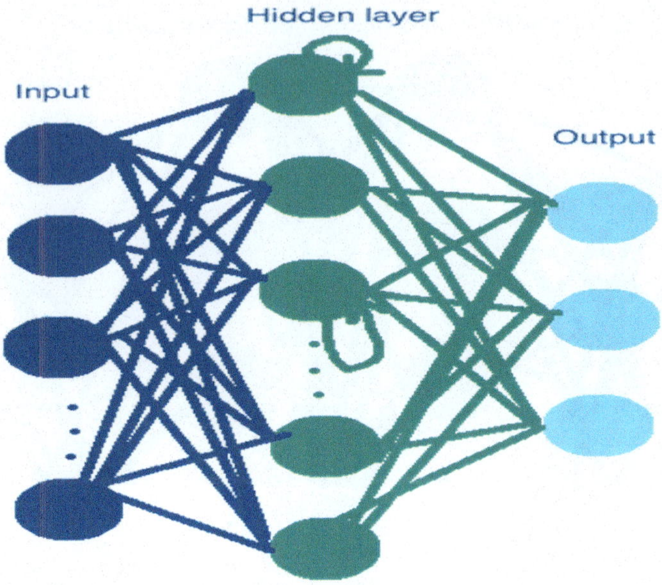


**Fig. 6.17** Recurrent neural network

**Fig. 6.18** Single layer NN architecture

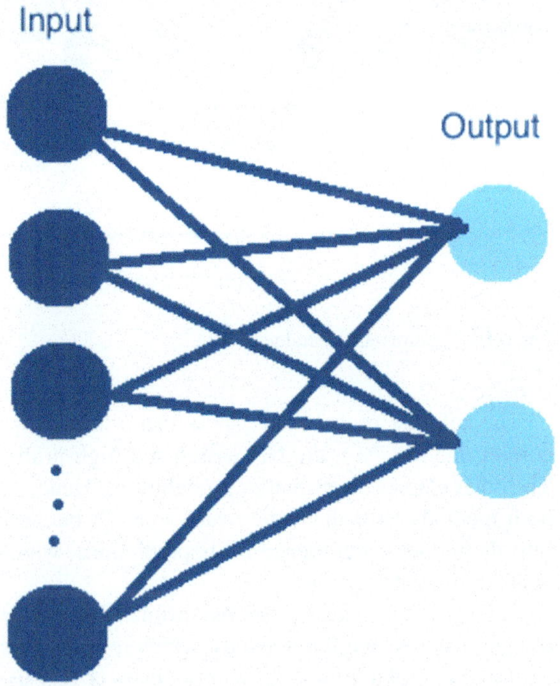

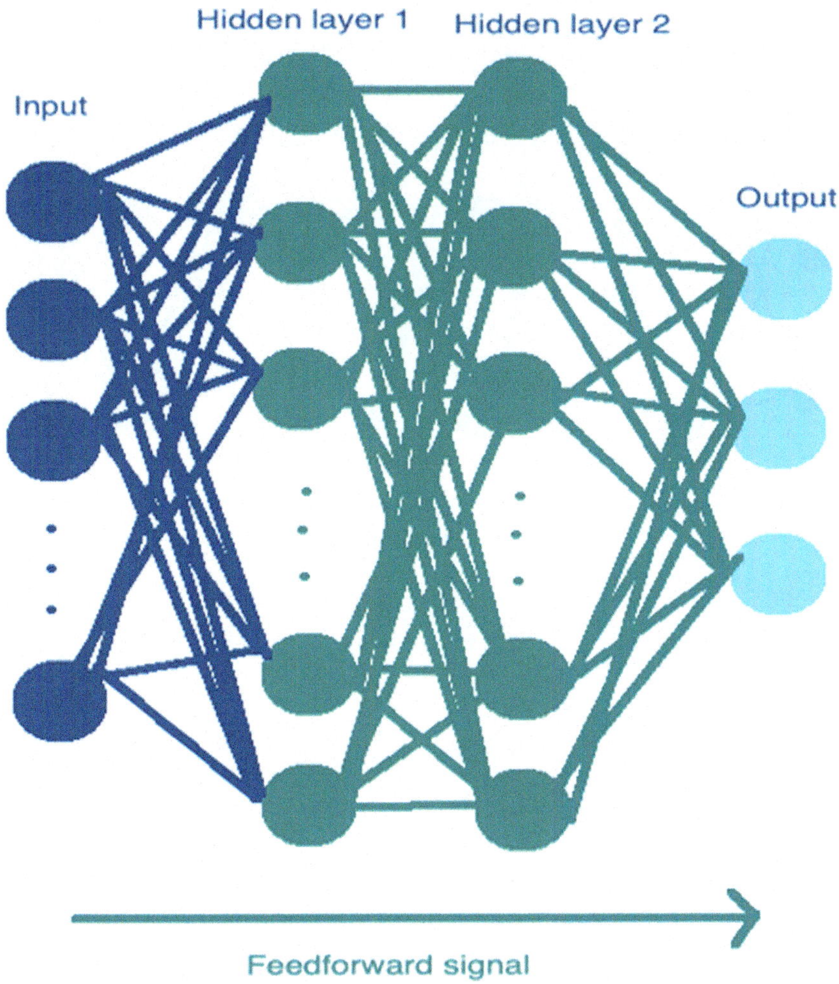


**Fig. 6.19** Multi-layer perceptron NN architecture

that illustrates how one class dominates the prediction when its score is much higher than other. The peak represents the confidence on the model on that specific class.

So far, we have addressed the NN's architecture, the NN's activation functions, but one still remains a mystery. How do NN learn? Is it by magic? No, once again it is only mathematics. Learning for an NN means tuning the weights, and other parameters. There are multiple learning paradigms, and we shall discuss some of them. The most known and used training algorithm is the *backpropagation.*

The backpropagation algorithm was developed in 1960, but it became known after it had been promoted by Rumelhard, Hinton, and Williams in 1989, [3]. The algorithm's idea is to compute the error of the network as the difference between

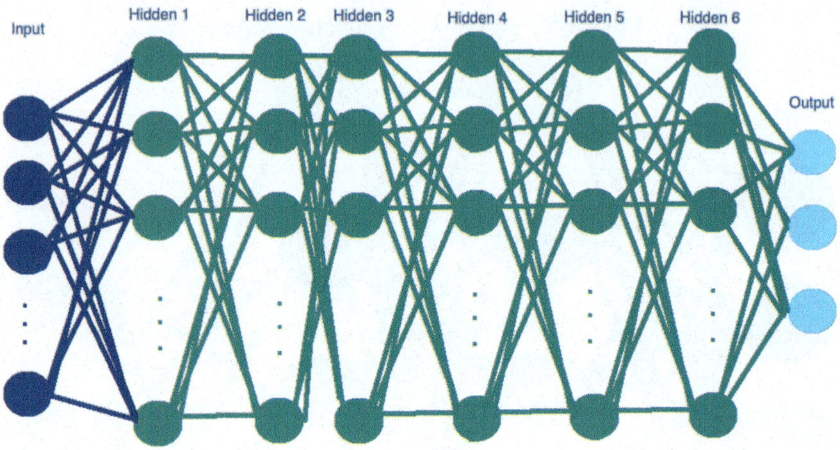


**Fig. 6.20** Deep learning architecture

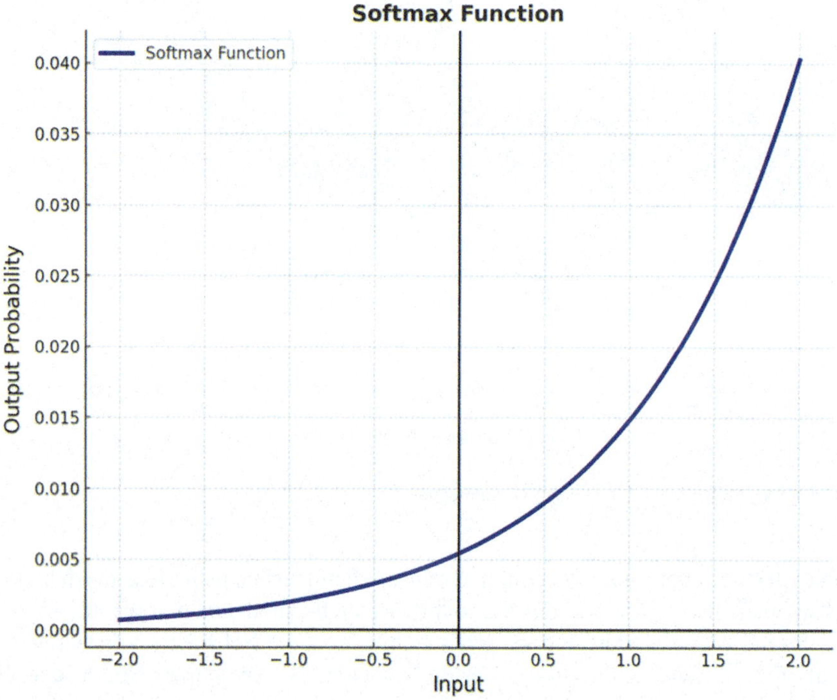


**Fig. 6.21** Softmax activation function plot

the ground truth and the network's output, and then propagate it back throughout the network until it reaches the input. Using the *chain rule* method, we adjust the weights and biases. The difference between the ground truth and network output represents the *loss* or the *error function*, and it is denoted by $L$. Since $L$ represents the error, in order for our network to work properly, we need to minimize it. First, we need to write the loss under a mathematical form. Let us denote $y_i$ as the network's output, and $d_i$ as the ground truth. Then, we have:

$$L = \frac{1}{n}\sum_i |y_i - d_i|^2,$$

We have discussed different minimization methods that are based on swarm intelligence, but now we will discuss the mathematical method called the *gradient descent*. This method is able to indicate the direction that you need to follow in order to optimize a function. Since we need to minimize the loss function our direction is the abrupt descent. During this abrupt descent, we need a new parameter named *learning rate*. The value of the learning rate establishes the size of the step during the abrupt descent. We can explain this descent with the following example: hiking down a mountain during the night, while suffering from amnesia (Fig. 6.22). Sounds terrible, isn't it?

Our only chance is to hike into the valley, to the nearest village. How do we know if we are hiking in the right direction? Using our feet we can determine if we are descending or ascending, hence we can determine the path by the direction of our boots. This is what we call the *slope*. Another issue that needs to be solved is the size of our steps. If they are too small, we might never get to the village in time, if they are too large, we might fall into the abyss. The size of our steps represents the learning rate.

In Fig. 6.23 we present how the graph looks when we are using a large learning rate, while in Fig. 6.24 we present a graph with a small learning rate.

In Fig. 6.23, we present the gradient decent with a large learning rate. The red connected points represent the steps that we take during our descent. The cost function cannot be seen in this graph because the large learning rate causes the gradient descent to move so far away from minimum value, so it does not appear in the graph. We can make the following key observations:

- The learning rate causes the gradient descent to miss the minimum, and just oscillate and diverge from it.
- As the large learning step makes the gradient jump back and forth, the cost increases.

In Fig. 6.24, we illustrate the gradient descent with a small learning rate. The cost function, depicted as the blue curve, can be spotted. The red connected dots are the steps the gradient takes during the descent. We can make the following key observations:

**Fig. 6.22** Gradient descent parallel to real life (generated with ChatGPT 4.0)

- Having a small learning rate allows us to take small and gradual steps, that allow our gradient descent to move steadily towards the minimum of the cost function, without missing it.
- Having a small learning rate permits us to avoid the oscillations that we saw when using a large learning rate, but it converges in a too slow process.

In this case we are facing a trade-off: either we choose a small learning rate and have a stable convergence at the cost of speed, or we choose a large learning rate but risk instability and divergence. Can we find the right balance between them? Can we optimize the gradient's descent performance?

Yes, we can, with the use of a dynamically adjusted learning rate. Figure 6.25 presents the graph of a gradient descent with dynamically adjusted learning rate.

From Fig. 6.25, we can see that the learning rate starts with a relatively large value and decreases over time with each step of the descent. This process allows the

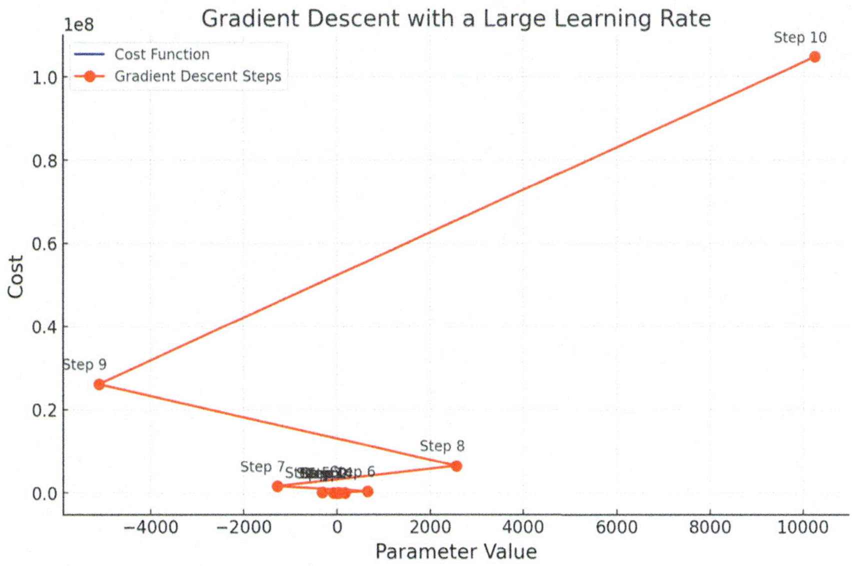


Fig. 6.23 Gradient descent with a large learning rate

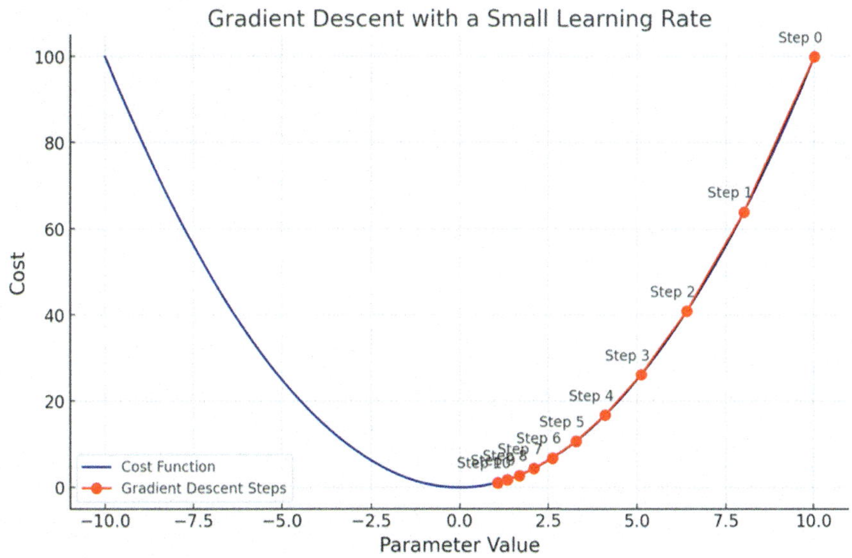


Fig. 6.24 Gradient descent with a small learning rate

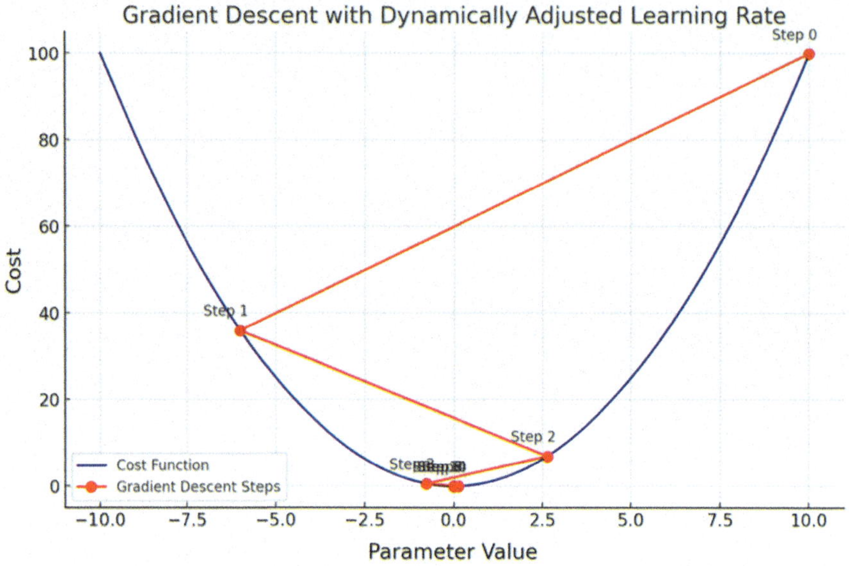


**Fig. 6.25** Gradient descent with a dynamically adjusted learning rate

gradient to take initial large steps and then gradually refine them as it gets closer to the minimum. We can make the following key observations:

- Initially, the learning rate oscillates similar to using the large learning rate.
- As the learning rate decreases, the gradient steps become smaller and thus more controlled, allowing the gradient to reach the minimum steadily and accurately.

Using a dynamically adjusted learning rate we find the best trade-off between speed and stability.

In Figs. 6.26 and 6.27 we show how the algorithm converges using a constant learning rate.

In Fig. 6.27 we present a contour plot of the optimization process using a constant learning rate. The blue line is the path that the algorithm takes as it converges towards the minimum, whereas the circles indicate the steps it makes along the way. The cost function is represented by the contour lines, and the center represents the optimal point that needs to be reached.

There are multiple ways to decrease dynamically the learning rate. One approach is by trying to lower its value epoch by epoch. This approach is known under the name of *learning rate decay*. Let us see now how these above plots look like if we use a decaying learning rate (Figs. 6.28 and 6.29).

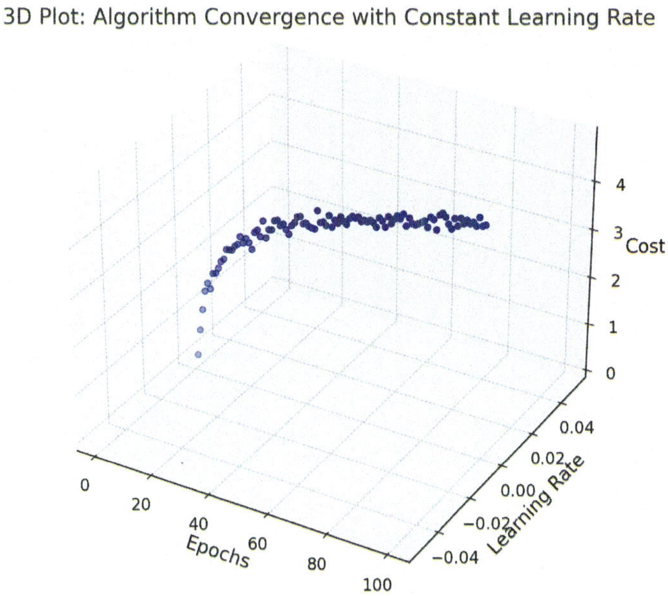


**Fig. 6.26** 3D plot of algorithm's convergence using a constant learning rate

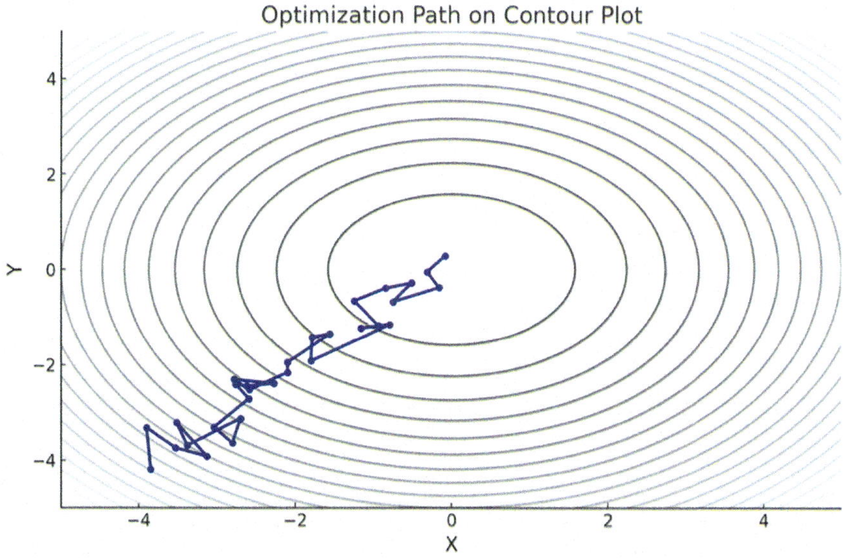


**Fig. 6.27** Algorithm converging with a constant rate

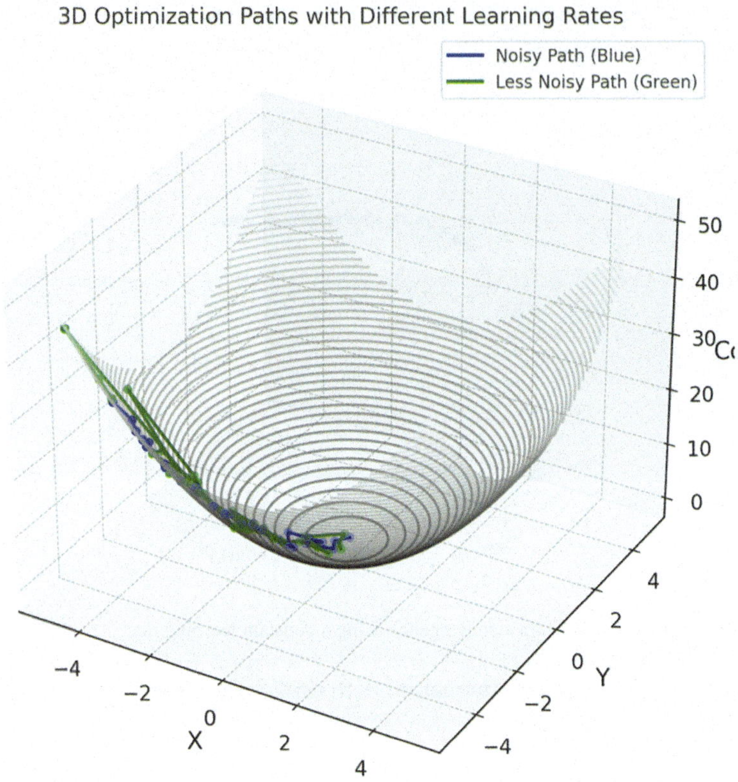


**Fig. 6.28** 3D optimization paths with constant vs decaying learning rate

The most common methods of lowering the learning rate are:

- The *step decay* in which we reduce the learning rate's value by a factor (e.g. divided by half).
- The *exponential decay* in which we decrease the learning rate's value epoch by epoch using an exponential schedule.
- The *time-based* in which we decrease the learning rate using the following formula:

$$learning_{rate} = \frac{initial\ learning\ rate}{1 + decay_{rate} \times epoch}.$$

Another way to dynamically adjust the learning rate is by using *learning rate schedulers.* This type of method decreases the learning rate using different criteria (i.e. obtained performance on the validation set). The most common schedulers are:

- The *ReduceLROnPlateau* scheduler reduces the learning rate when a predefined metric, such as the validation loss, does not improve anymore.

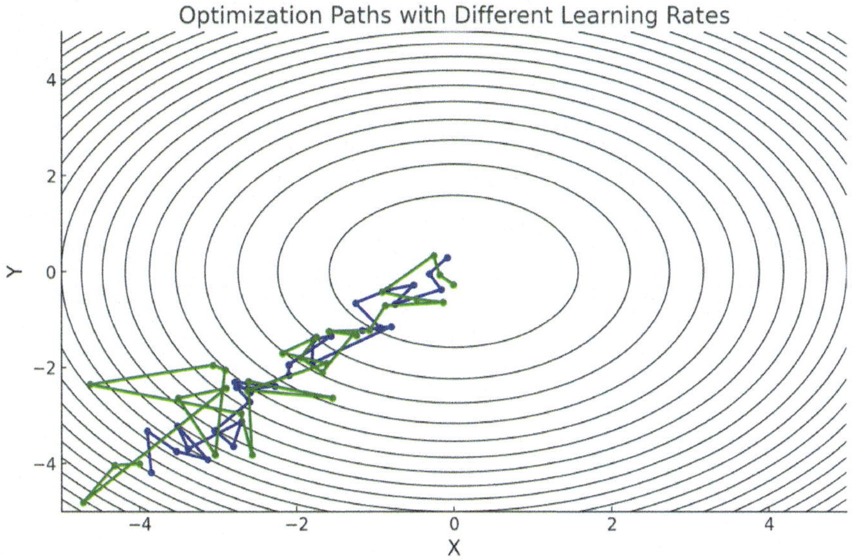


**Fig. 6.29** Algorithm converging with decaying learning rate versus constant learning rate

- The *Cyclic Learning Rates* scheduler alternates high and low values for the learning rate, values taken from a predefined range. In this way the NN can escape from local minima.

Let us plot these methods to see how they dynamically adjust the learning rates.

From Fig. 6.30, we can see that step decay (the blue plot) decreases the learning rate using a fixed factor of 30 epochs. The exponential decay (the green plot) decreases the learning rate exponentially, which leads to a smooth continuous decay. The cosine annealing (the red plot) gradually decreases the learning rate, while the cycle learning rate (the purple plot) oscillates between the minimum and the maximum value, thus enabling the model to explore different areas of the loss surface. The ReduceLROn-Plateau (the orange plot) used in this case the validation loss as metric and stopped improving for 20 epochs, hence it reduced the value of the learning rate. In this plot we can see how the model adapts to the model's performance during training.

Another approach for dynamically adjusting the learning rate is by using *adaptive learning rate*. In this type of approach, the learning rate is automatically adjusted for each parameter during training. The results of this approach surpass others due to their adaptability to the loss surface characteristics. The most common adaptive methods are:

- The *AdaGrad* method increases the learning rate for less frequent parameters, while decreasing it for the more frequent ones.
- The *RMSProp* method is a combination of AdaGrad with a moving average of squared gradients that prevent the learning rate to decrease too quickly.

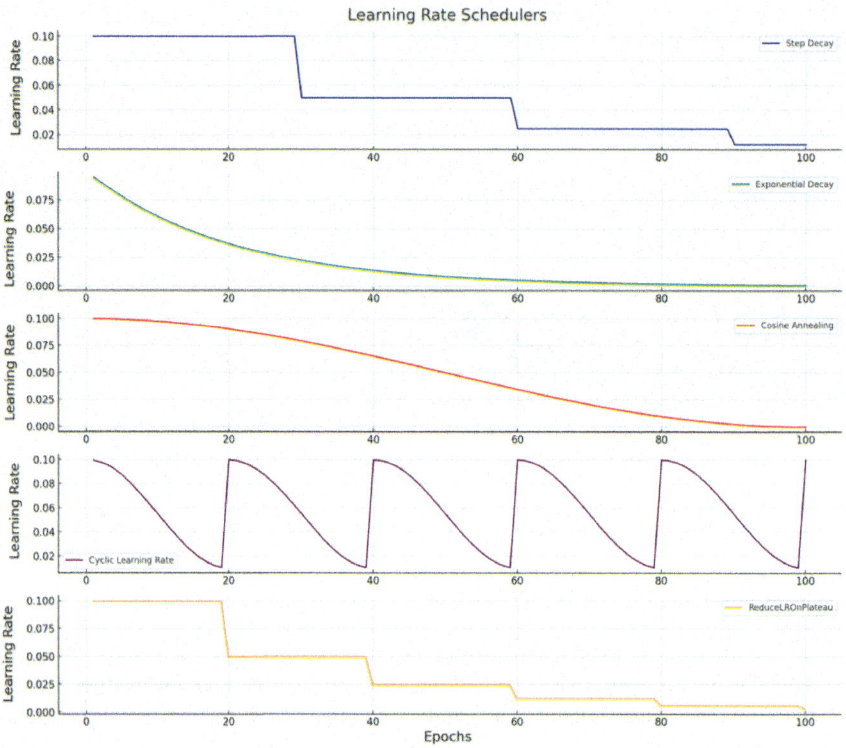


**Fig. 6.30** Dynamically adjustment of the learning rate plot comparison

- The *Adam* method adapts the learning rate for each parameter while using the momentum to accelerate the convergence.
- The *AdaDelta* method extends AdaGrad and adjusts the learning rate using a moving window of gradient updates.

In Fig. 6.31, we present the plots of how the learning rate adapts over time using different adaptive methods.

From Fig. 6.31, we can see that AdaGrad (the blue plot) starts with a high learning rate that quickly decreases as more and more gradients are accumulated. Why is that? AdaGrad balances the learning rate using the sum of the squared gradients, making the learning rate smaller as training progresses. You should use AdaGrad if your data is sparse. RMSProp (the green plot) keeps the learning rate more stable than AdaGrad, avoiding that rapid decay, due to the use of the moving average of the squared gradients. You should use RMSProp when dealing with non-convex optimization problems. Adam (the red plot) starts with a spike, after which it is able to stabilize it, while AdaDelta (the purple plot) adapts the learning rate more smoothly, remaining stable and avoiding a rapid decay.

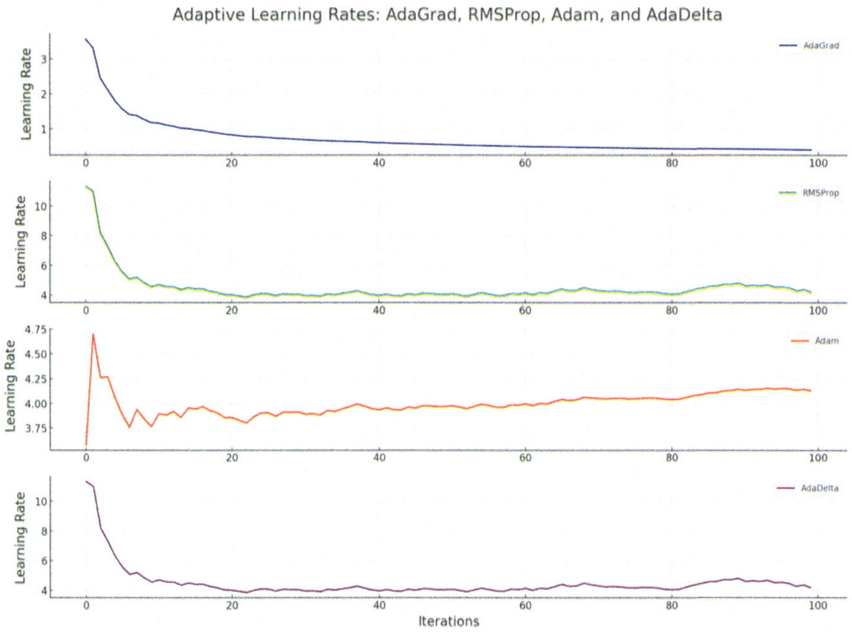


**Fig. 6.31** Adaptive learning rates plots

Another way to dynamically adjust the learning rate is by using the *cosine annealing*, also known as the *warm restarts*. In this type of method, we reset the learning rate periodically using a cosine schedule. The model escapes from local minima by cyclically increasing the learning rate, followed by regular reductions.

If we do not wish to adjust directly the learning rate, we could opt for *gradient clipping* which sets a threshold for the gradient's value during backpropagation. If the gradient surpasses that threshold, it is scaled down to the maximum allowed value. In this way we prevent the gradient from exploding.

You might wonder how we represent the gradient. Well, the gradient is an array of derivatives or slopes. The loss is computed using a continuous and differentiable function of the weights. The formula for computing the vanilla gradient (the standard) is:

$$\nabla L = \left( \frac{\partial L}{\partial w_1}, \ldots, \frac{\partial L}{\partial w_n} \right).$$

We update the synaptic weights using the learning rate, as:

$$\Delta w_i = -\eta \frac{\partial L}{\partial w_i}.$$

There are multiple learning paradigms that we can use to train the NNs. An alternative to the backpropagation algorithm is updating the weights using a swarm intelligence technique such as genetic algorithms, [4]. In this approach, the weights are encoded in a chromosome which represents a potential solution. By applying the operators described in Chap. 5, we are able to optimize the weights considering as fitness score the network's performance obtained using the weights from the chromosome. The error is computed as $chromosome_{error} = \frac{100 - fitness\_score}{100}$.

Another option for training NNs is by using Bayes' theorem, [5]. In the beginning, the weights are initialized using the rank correlation Goodman–Kruskal Gamma between the input and the output, $\Gamma(X_i, Y_i), i = 1, 2, \ldots, p, j = 1, 2, \ldots, q$, $p$ the number of features, and $q$ the decision classes. From a probabilistic point of view, we can consider the weights as being the values of a random variable $W_{ij}$. The conditional probability $P(E|w_{ij})$ can be considered the likelihood as well as the network's error.

If we assume that the events (features) $A_{ij}$ that correspond to $W_{ij}$ are a partition of the weight space, $W$, and denote by $E(n)$ the network's error at step $n$, then we can use the total probability's formula to compute:

$$P(E(n)) = \sum_{ij} P\left(E(n)|A_{ij}\right) \cdot P\left(A_{ij}\right),$$

with $i = 1, 2, \ldots, p, j = 1, 2, \ldots, q$.

The weights encode how strong is the liaison between the network's neurons. If we have a strong liaison, then we will have a high probability. Since, we presume that the weights can be considered as probabilities, then we can update them using Bayes' rule:

$$w_{ij}(n+1) = P\left(A_{ij}|E(n)\right) = \frac{P\left(E(n)|A_{ij}\right) \cdot P\left(A_{ij}\right)}{\sum_{ij} P\left(E(n)|A_{ij}\right) \cdot P\left(A_{ij}\right)},$$

having $P(E(n))$ the evidence, $P\left(A_{ij}\right)$ the prior probability, and $P\left(E(n)|A_{ij}\right)$ the likelihood, with $i = 1, 2, ..., p, j = 1, 2, \ldots, q$.

Therefore, we can rewrite the likelihood formula making use of the Goodman–Kruskal correlation rank between each feature and error at step $n$ as:

$$P\left(E(n)|A_{ij}\right) = \Gamma(X_i, E(n)), i = 1, 2, \ldots, p, j = 1, 2, \ldots, q.$$

More details on *objective versus subjective* dilemma can be found in [6].

Other interesting approaches for training single hidden layer feedforward networks are by using simple logistic regression for two-class decision problems, [7], or cascaded or parallel logistic regression for multiple decision problems, [8].

*Deep Learning*

It seems like nowadays many people confuse AI with Deep Learning (DL), which is a big mistake! DL is just a broad term that encompasses all types of deep neural networks. A specific type of such networks is the Convolutional Neural Network

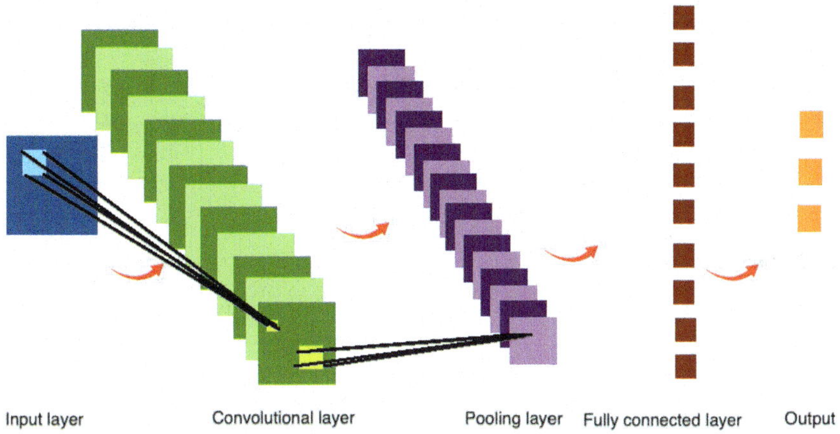


**Fig. 6.32** The CNN's architecture

(CNN). Another big mistake is considering that DL and CNNs are something new. They are, if you consider a technique developed back in the 40s new. In 1979, Kunihiko Fukushima developed the neocognitron, which is a neural network that had a lot of layers displayed in a hierarchical manner, [9]. In 1989, Yan LeCun presented a new type of CNN which was able to recognize handwritten digits. The CNN was trained using the backpropagation algorithm, [10].

Fifteen years after the first AI winter ended, a second one began. NNs and DL were forgotten. Nobody studied this area. AI became a pseudoscience. Due to video games, graphics processing units (GPUs) were being upgraded. This led to computers becoming much faster, which was the perfect scenario for DL to rebirth like the Phoenix bird in 1999. Two years later, the revolution of Big Data began. Data became 3D. The ImageNet free dataset was released in 2009, due to Prof. Fei-Fei Li and her team at Stanford. ImageNet contains 14 million labeled images that can used to train CNNs.

CNNs are designed to process data that is structured in a grid-like manner. That is why they are perfect for processing images and video. In a CNN architecture we can encounter the following types of layers (Fig. 6.32):

- The *input* layer: CNNs generally take as input images, which are represented by a matrix of pixel values. If the image is a color one, we have three dimensions that represent the color channels (red, green, blue), the height, and the width of the image. If the image is in grayscale, then we have only two dimensions, the height, and the width.
- The *convolutional* layer: in this layer the magic of CNN happens.
    - The core operation in a CNN is the *convolution operation*, in which we have a *filter* or a *kernel* (a small matrix that contains weights) that is applied to our input data. The filter moves, slides, or convolves across the input data, while

performing element-wise multiplication and summation. The result is a feature map.

- The outputs of the convolution operation are different feature maps that represent different features (edges, complex patterns, blobs of color, textures, etc.) in diverse location in the input image.
- In general, in each convolutional layer we apply *filters* to extract different features, hence we have multiple feature maps.

- *Activation functions*:
  - Just like in the case of classical NNs, in CNNs we have activation functions after each convolution. This provides non-linearity to the model. In general, the most used activation function is ReLU.
  - The non-linearity characteristic helps the CNN model complex patterns in data, because it can learn non-linear relationships between the input data and the output.

- The *pooling layer*:
  - Pooling layers are used for reducing the spatial dimensions (i.e. height and width) of the feature maps. By down sampling them, we reduce the number of parameters, and hence the computational load. The model becomes more robust to small translations in the input data.
  - The most used pooling operation is *max pooling*, [11]. In this type of pooling, we are taking the maximum value from each region of the feature map. Another type of pooling operation is the *average pooling*, in which we compute the average value of each region.

- The *fully connected layers* or *dense layers*: after multiple convolutions and pooling layers, the feature maps are *flattened* into a single 1D array, that is fed to the next fully connected layer. In general, the dense layers can be found at the end of the CNN. They are used to make predictions and classifications. All the neurons in these types of layers are connected to all the neurons in the previous layers.
- The *output layer* is the final layer that return the CNNs output. It has a softmax activation function applied to it, if we are using it for classification purposes, or a linear activation function, if we are using it for regression purposes.

In a convolutional layer, the neurons are not connected to all the neurons in the previous layer, they are connected only to a local region of the input, known as the *receptive field*. Through these receptive fields, the CNN can focus on localized features, which might come in handy when dealing with images.

Another important concept in a CNN is the *stride*. The stride expresses how much does the filter move when it convolves throughout the input. If a stride is equal to 1, then the filter moves one pixel at a time, if the stride is 2, it moves two pixels at a time, and so on. When choosing the value for the stride, we must find the trade-off between speed and losing details. A large value of the stride means a smaller size of the feature map, which leads to fast computations, but also to possibly losing important details.

Sometimes important patterns and features are found at the border of the input image. Because we do not want to lose that information, we need to add padding to the input. Padding means adding extra pixels around the border of the image, before applying the convolution operation.

Let's see how this works. Let's presume that we have an input matrix of $4 \times 4$, a kernel of $2 \times 2$, padding $= 1$, and stride $= 1$. Our input is:

$$Input = \begin{bmatrix} 1\,2\,3\,0 \\ 4\,5\,6\,1 \\ 7\,8\,9\,0 \\ 1\,2\,3\,4 \end{bmatrix}.$$

and kernel:

$$kernel = \begin{bmatrix} 1\,0 \\ 0\,1 \end{bmatrix}.$$

Let us apply the padding:

$$padded\ input = \begin{bmatrix} 0\,0\,0\,0\,0\,0 \\ 0\,1\,2\,3\,0\,0 \\ 0\,4\,5\,6\,1\,0 \\ 0\,7\,8\,9\,0\,0 \\ 0\,1\,2\,3\,4\,0 \\ 0\,0\,0\,0\,0\,0 \end{bmatrix}.$$

Now to compute the convolution. As the kernel slides over the padded input matrix, we perform an element-wise multiplication followed by a summation of the results. Thus, we produce the feature map. For an easier understanding, we shall present the computations for computing two elements:

$$(1 \times 1) + (2 \times 0) + (4 \times 0) + (5 \times 1) = 1 + 0 + 0 + 5 = 6$$

$$(2 \times 1) + (3 \times 0) + (5 \times 0) + (6 \times 1) = 2 + 0 + 0 + 6 = 8$$

Following this example, we obtain the feature map:

$$feature\ map = \begin{bmatrix} 6\,8\,4\,1 \\ 9\,1\,9\,1 \\ 8\,1\,9\,4 \\ 2\,5\,7\,4 \end{bmatrix}.$$

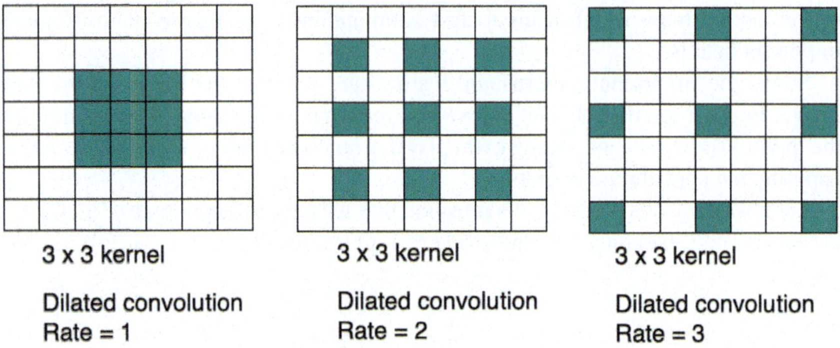


**Fig. 6.33** Example of dilated convolutions

If we apply max pooling to this feature map we obtain:

$$\max \textit{pooling} = \begin{bmatrix} 1\ 4\ 9 \\ 1\ 4\ 9 \end{bmatrix},$$

whereas if we apply average pooling to the same feature map, we obtain:

$$\textit{average pooling} = \begin{bmatrix} 9.25 & 3.75 \\ 8.5 & 6 \end{bmatrix}.$$

In 2016, Yu and Koltun developed a *dilated convolution*, [12]. If we add spaces between a filter's cell, we obtain a dilation. A dilated convolution is also known as an *atrous convolution* (Fig. 6.33). By using dilations, we are able to increase the receptive field, without increasing the number of parameters, hence we keep the same amount of computations. Dilated convolutions are especially useful when we are performing image segmentations, object detection, or sequence modeling.

Not all CNNs architectures use fully connected layers before the output layer. For instance, in 2016, Goodfellow and his colleagues proposed the *sparse connectivity* or *sparse weights*, [13]. The kernel size controls the sparse connectivity. By using a small kernel, we are able to detect small details.

CNNs can be trained the same way as classical NNs.

CNNs can assist clinicians providing automated tools for diagnosis, prognosis, and tailored treatment planning. They can classify images into different categories, for example detecting pneumonia or COVID-19 from X-rays, or cancerous lesions in a mammogram. Through segmentation they can highlight abnormalities in images, quantify their size, shape, and volume of the pathology. For instance, we can use CNNs to segment brain tumors in MRI images. Besides detecting abnormalities, we can use CNNs to localize specific features, for instance polyps in colonoscopies or nodules in lung CT scans. In pathology, we can use CNNs to identify and classify cancerous cells, grade tumors, and ultimately predict the patient's outcome based on those tissue samples.

In this chapter we have covered the theory behind neural networks and how they can be used in improving the radiology and pathology department. In the next chapter, we will discover how to improve the surgical department with the use of AI.

## References


1. McCulloch, W., Pitts, W.: A logical calculus of the ideas immanent in nervous activity. Bull. Math. Biophys. **5**, 115–133 (1943)
2. Hebb, D.O.: The Organization of Behavior. A Neuropsychological Theory. Wiley, New York (1949)
3. Rumelhart, D.E., Hinton, G.E., Williams, R.J.: Learning representations by backpropagating errors. Nature **323**, 533–536 (1986)
4. Belciug, S., Gorunescu, F.: A hybrid neural network/genetic algorithm system applied to the breast cancer detection and recurrence. Exp. Syst. J. Knowl. Eng. **30**(3), 243–254 (2013)
5. Belciug, S., Gorunescu, F.: Learning a single-hidden layer feedforward neural network using rank correlation-based strategy with application to high-dimensional gene expression and proteomic spectra data sets in cancer detection. J. Biomed. Inf. **83**, 159–166 (2018). https://doi.org/10.1016/j.jbi.2018.06.003
6. Wagenmakers, E.-J. Lee, M., Lodewckx, T., Iverson, G.: Bayesian evaluation of informative hypothesis (statistics for social and behavioral science). In: Hoijtink, H., Klugkist, I., Boelen, P. (eds.) Bayesian Versus Frequentist Interference, pp. 181–207. Springer (2008)
7. Belciug, S.: Logistic regression paradigm for training a single-hidden layer feedforward neural network. Application to gene expression datasets for cancer research. J. Biomed. Inform. **102**, 103372 (2020). https://doi.org/10.1016/j.jbi.2019.103372
8. Belciug, S.: Parallel versus cascaded logistic regression trained single-hidden feedforward neural network for medical data. Exp. Syst. Appl. **170**, 114538 (2021)
9. Fukushima, K.: Neocognitron: a self-organizing neural network model for a mechanism of pattern recognition unaffected by shift in position. Biol. Cybern. **36**(3), 193–202 (1980)
10. LeCun, Y.: Generalization and Network Design Strategies (Technical Report, CRG-TR_89-4). University of Toronto (1989)
11. Zhou, Y., Chellappa, R.: Computation of optical flow using a neural network. In: IEEE International Conference on Neural Networks, pp. 71–78 (1988)
12. Yu, F., Koltun, V.: Multi-scale context aggregation by dilated convolutions. Comput. Vis. Pattern Recogn. (2016). arvix.org/abs/1511/07122
13. Goodfellow, I., Bengio, Y., Courville, A.: Deep Learning. MIT Press, Cambridge, MA/London (2016)

# Chapter 7
# The Operation Theatre Becoming Smart

**Abstract** As surgeons prepare for an operation, the operation theater must keep up with them. No matter how many imaging tests, blood tests you perform prior to surgery, when it comes to opening up a patient, surprises appear instantly. Hence, we like to explore the possibility of making the operation theater smart, so we give the surgeon, the anesthetists, nurses, and patient, the best outcome possible.

Preparing for a surgery takes a lot of time. To improve the outcomes, in recent years there have been several attempts to integrate AI in surgery. Prior to a surgery, the patient undergoes different tests, whether they are imaging test or blood tests. These tests produce numerous data, that if combined properly can lead to a smarter operating theatre. For instance, an Artificial Intelligence tool based on Gaussian process regression, time warping, and support vector machines were able to predict post-operative surgical site infections for gastrointestinal surgeries, [1]. Surgical site infections are very common, representing 30% of all acquired infections during a hospital stay, [2, 3]. If a patient gets such an infection, she/he/they might have a prolonged stay in the hospital, up to 10 days, her/his/their quality of life might decrease, and the mortality might increase by 3%. Each readmission into the hospital due to a surgical site infection costs the hospital around $27,000, [4].

Depending on the gravity of the surgical site infection, the doctors apply different treatment schemes. If it is a superficial infection, it can be easily treated with oral antibiotics and surgical debridement, otherwise, if it is a deep infection, the treatment requires intravenous antibiotics, drainage, and laparotomies. Blood tests can be used as predictive features. They are always performed prior to surgery, and they do not pose to much burden on the patient. It is possible to predict the patient trajectory after surgery using blood tests, [5].

Neural Networks (NNs) have been able to predict pancreatitis severity in six hours after admission, better than acute physiology and chronic health evaluation II's, by at least 10%, [6]. Patient history, treatment, length of stay, and blood pressure were the features that had been fed to a NN to predict the in-hospital mortality following open abdominal aortic aneurysm. The reported results show 87% sensitivity, 96.1% specificity, and overall accuracy 95.4%.



S. Belciug, *The Hospital Manager's Guide to Artificial Intelligence*, Intelligent Systems Reference Library 263, https://doi.org/10.1007/978-3-031-80314-7_7

The surgical operating theatre has adopted in recent years the concept of Internet of Things (IoT). This subarea used in surgery is called Internet of Surgical Things (IoST) and involves software, smart surgical instruments, and sensors. The key components of the IoST are:

- *Smart Surgical Instruments* that have sensors embedded in them, sensors that monitor and report data in real-time. Some smart surgical instruments are the smart scalpel, smart forceps, and different suturing devices. All these instruments give valuable insights and feedback to the surgeons during the procedures.
- *Robotic Surgical Systems* (i.e. da Vinci robot) are connected to the IoST network. These robots are able to perform minimally invasive surgery having a higher precision than a human surgeon.
- *Wearable and Implantable Devices* (i.e. patches, smart watches) monitor non-stop the patient's vital signs: heart rate, blood pressure, oxygen levels, during and after surgery. Some of these devices can be implanted in the patient to monitor internal conditions (i.e. intracranial pressure). The recovery of the patient is tracked, and any potential early sign of complication is detected, and the device sends the doctor an alert if any kind of intervention is needed.

The data from these sensors is collected and then fed to different AI algorithms that analyze it in real time and assist in decision-making during surgery (Fig. 7.1).

Virtual Reality (VR) and Augmented Reality (AR) are not just for video games. They are used in surgery to display interactive 2D and 3D images of the patient's anatomy or physical procedures. In this way the efficiency is maximized, and the risk is reduced.

VR can provide training by allowing the surgeon to practice a priori the procedures in a risk-free environment. By practicing multiple times, the surgeons can refine their techniques minimizing the patient's risk. Using VR, different surgical scenarios can be simulated, and hence the surgeon gains more experience and sees how she/he/they reacts under pressure when unforeseen complications appear. Please keep in mind, that this doesn't mean that the experience gained with VR surgeries can replace the experience gained with real surgeries.

Some VR environments have haptic feedback that give the sense of touch and resistance of tissue and instruments. VR enables remote collaboration between surgeons. In a case where the surgeon needs help, guidance and advice, another, more experienced surgeon can virtually join the surgery. Besides this, the VR interface can display the patient's vital signs, saving time and conservating focus, because the surgeon doesn't need to look away from the surgical site to check the vitals.

The equipment needed for VR surgery is:

- VR head-mounted display that allows the surgeon to immerse in the virtual environment. It is able to track the head movements so that the view is adjusted correctly.
- Haptic gloves or haptic controllers that provide the tactile feedback. The simulation must feel real to the surgeon in terms of tissues, instruments, etc.

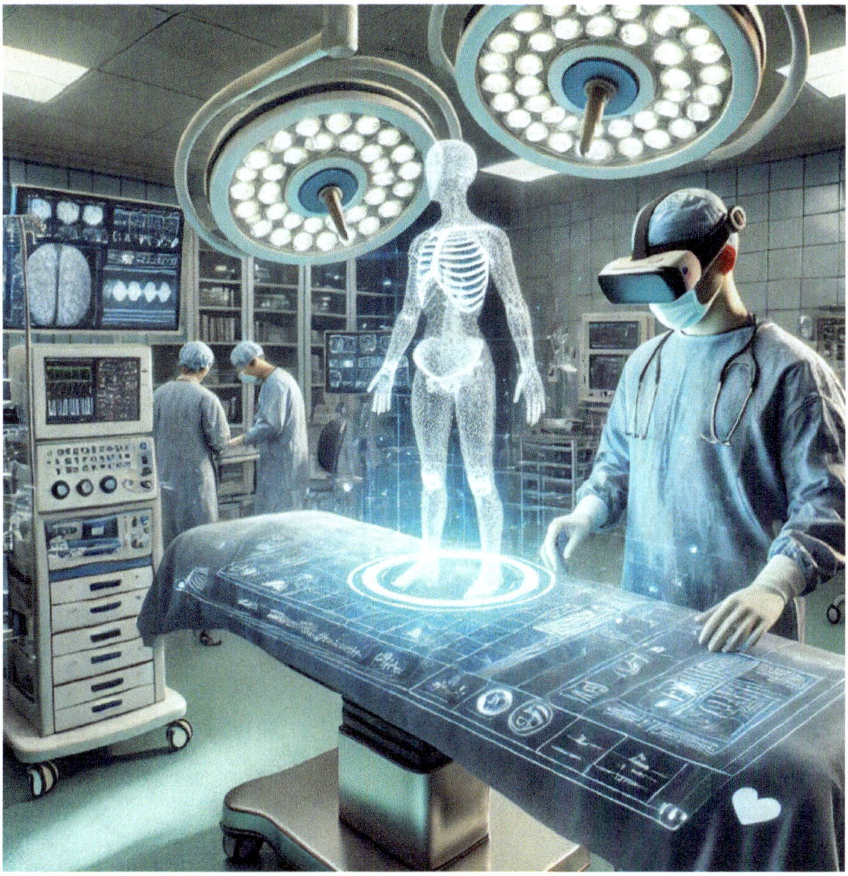

**Fig. 7.1** Virtual reality in surgery (image generated with ChatGPT 4.0)

- Tracking systems to track the movement of the user so that virtual interaction is realist and also responsive.
- Powerful computers with high-end GPUs that are able to render realistic 3D environment and are able to manage real-time interactions.
- VR software that is able to generate the virtual surgery, to provide human interaction, and collect data regarding the user's performance.
- Fully integrated VR surgery room simulators (Fig. 7.2).

AR is quite different from VR. AR overlays digital information onto the real patient in real-time. Sounds a little bit sci-fi, doesn't it? AR can overlay 3D models of the bones, organs, blood vessels, and thus the surgeon can make more precise incisions. If you recall, some puzzles give you the image that you need to create, and you just place the pieces of the puzzle onto it. Through AR, the surgeon can see inside the body using enhanced MRI, CT, or US scans, without making large incisions.

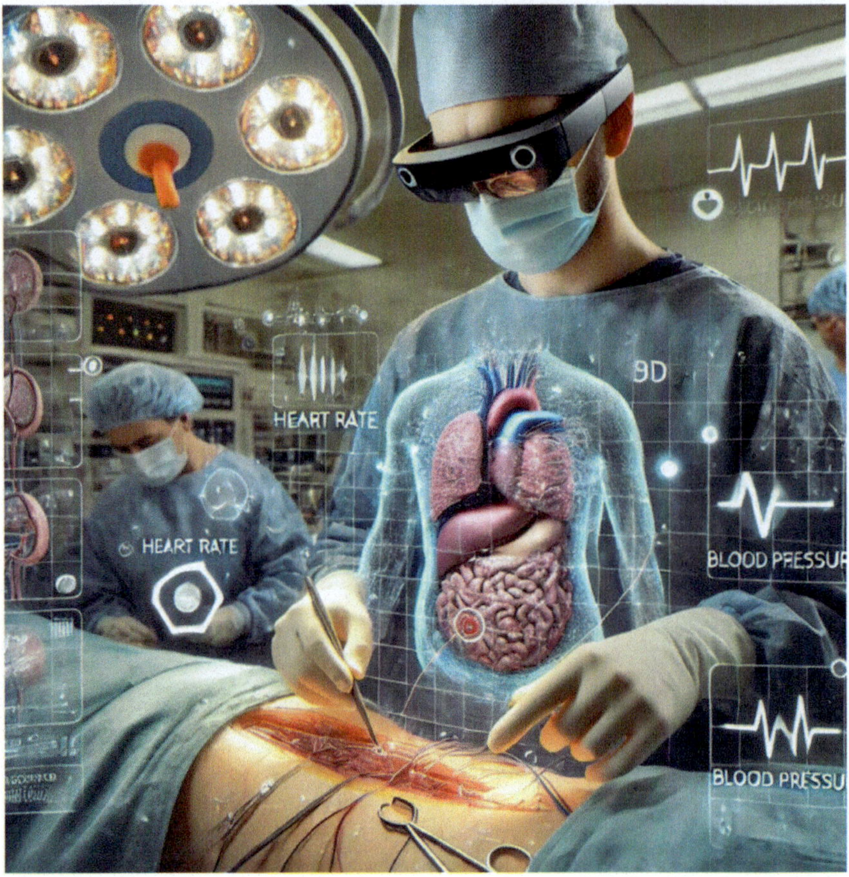


**Fig. 7.2** Augmented Reality in surgery (image generated with ChatGPT 4.0)

This suits better minimally invasive surgeries. Nevertheless, even if we advocate for bringing AI to the O.R., we still believe that it has its limits, and sometimes, the real surgical site provides better intel than the virtual one.

Another interesting and important aspect of using AR in the O.R. is the fact that it can place virtual markers on the patient's body, and thus guide the surgeon where to cut or place the instruments. Obviously, an experienced surgeon does not need all this, but for a young one this feature provides the best possible help. AR provides real-time feedback if a potential nerve or blood vessel might be damaged during surgery.

The basic components needed to have AR in the hospital's O.R. are:

- An AR display device that contains headsets, smart glasses, and sometimes AR-enabled monitors. The AR display device projects digital intel while maintaining the visibility of the physical world.

- A tracking system made of cameras and sensors that track the patient's position, the surgical tools' position, and the movements of the surgeon.
- Specialized AR software that is able to process imaging data (CT, MRI, US) and can convert it to 3D models that can be projected on the AR display device. The software must be connected to the patient's EHR.

For more than a decade now, VR has been used for endoscopic training, [7]. Using VR, it has been shown that the operative time is shortened, and the error rate minimized, [8]. VR was tested in preoperative training for young surgeons, and as a simulator for surgeries with great results, [9, 10]. Using VR there have been developed virtual dissection tables, which are a virtual representation of both male and female cadavers. They are considered a promising way of teaching surgeons, [11]. In what regards patients, it has been demonstrated that by using VR, they understand better prospective procedures, and thus, their preoperative anxiety is reduced, [12–14].

The innovation in VR/AR software and hardware is now widespread. It is expected to have a market growth of 50 billion USD by 2026, (https://www.idc.com/getdoc.jsp?containerId=prUS49916122—accessed August 30, 2024). New medical apps use HoloLens and dedicated software platforms in clinical settings, [15–24].

In general, the physical environment needed to use VR for education and training purposes is pricy, [25, 26]. However, during the past few years, multiple low-cost devices have appeared on the market. The cost of such a low-cost device does not exceed $500, therefore it can be acquired. We mention the following devices: Oculus Rift, HTC Vive, Pico, Quest, [26, 27]. For instance, in [27], the Oculus Rift was used in combination with Leap Motion sensor. Oculus uses a wired computer connection and through the Oculus software is able to display through screencast the 3D virtual environment of the Rift device. Leap Motion captures the user's motions and transforms them into commands for a computer or a VR device.

In order to create a VR environment, we need image scans and recorded in vivo procedures. Having these we can create 3D models of the internal organs. If needed, some important details of the 3D models can be printed, [28, 29]. In [27], the researchers were able to make a VR of an abdominal surgery using open-source software and low-cost hardware device. Using CT and MRI scans and a dedicated open-source software like OsiriX, they built a 3D model of the organs, [30]. With Osirix, once the 3D scene is built, we can turn on and off different organs, and they become transparent, hence we can see any details inside or below. This is a segmentation process which can be:

- Automatic: all the organs are recognized without interaction with a human. Not all body areas are available yet.
- Semiautomatic: the segmentation is done after the user sets different parameters.
- Manual: the user has a pen instrument, and it uses it to segment manually the organs.

In this study, the authors considered the pancreas, the liver and kidney for the VR surgery. The study included 15 oncological patients that needed surgery. The 15 patients were divided into three groups, 5 patients in the group scheduled for pancreatic surgery, 5 patients in the group scheduled for the liver surgery, and 5 patients in the group scheduled for the kidney surgery. CT scans were obtained from all patients, and contrast-enhanced MRI from two patients. The organ segmentation was semiautomatic and was performed using 3D Slicer, an open-source software developed by Harvard Medical School and the Massachusetts Institute of Technology (https://www.slicer.org—accessed September 1, 2024).

The 3D reconstruction was imported in Unity3D, which is a platform used to develop games, (https://unity3D.com—accessed September 1, 2024). Unity3D enables free integration with Oculus Rift and Leap Motion. The most important scenes were 3D printed. The printed models were validated by a radiologist. The study received the clinicians' valid appreciation in using VR for abdominal surgery.

Regarding AR, we mention the following AR apps for clinical care:

- AccuVein is an app that helps to locate the patient's veins accurately for blood withdrawal or injections (accuvein.com—accessed September 1, 2024).
- Brain Power is an app for teaching different life skills for children and adults on the autism spectrum (brain-power.com—accessed September 1, 2024).
- Eye Decide is an app that helps clinicians demonstrate how different conditions impede the patient's eyesight.
- Surgical Navigation Advanced Platform app allows surgeons lay out their surgical plans.
- Anatomy 4D app allows a details visualization of bone structures and organ systems when the device is pointed at special downloaded templates.

The benefits of using VR and AR in the operating theatre are obvious. We are still far from seeing this being implemented in every hospital from every country. The technology is yet expensive, even if some low-cost devices have appeared on the market. These devices are used for research purposes and not cleared to be used in regular surgeries. Nevertheless, step-by-step, we are approaching that era, where surgeries will become safer.

## References

1. Soguero-Ruiz, C., Fei, W.M., Jenssen, R., et al.: Data-driven temporal prediction of surgical site infection. AMIA Annu. Symp. Proc. 1164–1173 (2015)
2. Magill, S.S., Hellinger, W., Cohen, J., Kay, R., Bailey, C., Boland, B., et al.: Prevalence of healthcare-associated infections in acute care hospitals in Jacksonville, Florida. Infect. Control **33**(03), 283–291 (2012)
3. De Lissovoy, G., Fraeman, K., Hutchins, V., Murphy, D., Song, D., Vaughn, B.B.: Surgical site infection: incidence and impact on hospital utilization and treatment costs. Am. J. Infect. Control **37**(5), 387–397 (2009)
4. Owens, P.L., Barrett, M.L., Raetzman, S., Maggard-Gibbons, M., Steiner, C.A.: Surgical site infections following ambulatory surgery procedures. JAMA **311**(7), 709–716 (2014)
5. Skrovseth, S.O., Augestad, K.M., Ebadollahi, S.: Data-driven approach for assessing utility of medical tests using electronic medical records. J. Biomed. Inf. **53**, 270–276 (2015)

6. Monsalve-Torra, A., Ruiz-Fernandez, D., Marin-Alonso, O., Soriano-Paya, A., Camacho-Mackenzie, J., Carreno-Jaimes, M.: J. Biomed. Inf. **62**, 195–201 (2016)
7. Coleman, J., Nduka, C.C., Darzi, A.: Virtual reality and laparoscopic surgery. Br. J. Surg. **81**, 1709–1711 (1994)
8. Nagendran, M., Gurusamy, K.S., Aggarwak, R., et al.: Virtual reality training for surgical trainees in laparoscopic surgery. Coch. Database Sust. Rev. **8**, CD006675 (2013)
9. Cates, C.U., Lonn, L., Gallagher, A.G.: Prospective randomized and blinded comparison of proficiency-based progression full physics virtual reality simulator training versus invasive vascular experience for learning carotid artery angiography by very experienced operators. BMJ Simul. Technol. Enhanc. Learn. **2**, 1–5 (2016)
10. Ali, S., Qandeel, M., Ramakrishna, R., Yang, C.W.: Virtual simulation in enhancing procedural training for fluoroscopy-guided lumbar puncture. A pilot study. Acad. Radiol. **25**, 235–239 (2018)
11. Bianchi, S., Bernardi, S., Perilli, E., Cipollone, C., Di Biasi, J., Macchiarelli, G.: Evaluation of digital technologies during anatomy learning in nursing school. Appl. Sci. **10**, 2357 (2020)
12. Bekelis, K., Calnan, D., Simmons, N., MacKenzie, T.A., Kakoulides, G.: Effect of an immerse preoperative virtual reality experience on patient reported outcomes: a randomized controlled trial. Ann. Surg. **265**, 1068–1073 (2017)
13. Noben, L., Goosses, S., Truijens, S.E.M., Van Berckel, M.M.G., Perquin, C.W., Slooter, G.D., Van Rooijen, S.J.: A virtual reality video to improve information provision and reduce anxiety before caesarean delivery: randomized controlled trial. JMIR Ment. Health **6**, e15872 (2019)
14. Mahajan, U.V., Sunshine, K.S., Herring, E.Z., Labak, C.M., Wright, J.M., Smith, G.: Virtual reality in presurgical patient educationL a scoping review and recommended trial design guidelines. Am. J. Surg. **222**, 704–705 (2021)
15. Amini, S., Kersten-Oertel, M.: Augmented reality mastectomy surgical planning prototype using HoloLens template for healthcare technology letters. Healthc. Technol. Lett. **6**, 261–265 (2019)
16. Coelho, G., Rabelo, N.N., Vieira, E., Mendes, K., Zagatto, G., de Oliveria, R.S., Raposo-Amaral, C.E., Yoshida, M., et al.: Augmented reality and physical hybrid model simulation for preoperative planning of metopic craniosynostosis surgery. Neurosurg. Focus. **48**, E19 (2020)
17. Garcia-Mato, D., Moreta-Martinez, R., Garcia-Sevilla, M., Ochandiano, S., Garcia-Leal, R., Perez-Mananes, R., Calvo-Haro, J.A., Salmeron, J.I., Pascau, J.: Augmented reality visualization for craniosynostosis surgery. Comput. Methods Biomech. Biomed. Eng. Imaging Vis. **9**, 392–399 (2021)
18. Pratt, P., Ives, M., Lawton, G., Simmons, J., Radev, N., Spyropoulou, L., Amiras, D.: Through the HoloLens$^{TM}$ looking glass: augmented reality for extremity reconstruction surgery using 3D vascular models with perforating vessels. Eur. Radiol. Exp. **2**, 2 (2018)
19. Wesselius, T.S., Meulstee, J.W., Luijen, G., Xi, T., Maal, T.J.J., Ulrich, D.J.O.: Holographic augmented reality for DIEP flap harverst. Plast. Reconstr. Surg. **147**, 25E-29E (2020)
20. Giacalone, G., Yamatomo, T., Belva, F., Hayashi, A., Dori, Y., Zviman, M.M., Gysen, M., Nam, H.H., Jolley, M.A., Kato, M.: The application of virtual reality for preoperative planning of lymphovenous anastomosis in patient with a complex lymphatic malformation. J. Clin. Med. **8**, 8–371 (2019)
21. Cho, K.H., Papay, F.A., Yanof, J., West, K., Gharb, B.B., Rampazzo, A., Gastman, B., Schwarz, G.S.: Mixed reality and 3D printed models for planning and execution of face transplantation. Ann. Surg. **274**, E1238–E1246 (2021)
22. Mespreuve, M., Waked, K., Collard, B., De Ranter, J., Vanneste, F., Hendrickx, B.: The usefulness of magnet resonance angiography to analyze the variable arterial facial anatomy in effort to reduce filler-associated blindness: anatomical study and visualization through an Augmented Reality application. Aesthet. Surg. J. Open Forum, **3**, ojab018 (2021)
23. Waked, K., Mespreuve, M., De Ranter, J., Collard, B., Hahn, S., Hendrickx, B.: Visualizing the individual arterial anatomy of the face through augmented reality—a useful an accurate tool during dermal filler injections. Aesthet. Surg. J. Open Forum. **4**, ojac012 (2022)

24. Yaremenko, A.I., Lysenko, A.V., Ivanova, E.A., Galibin, O.V.: Augmented reality technology for auricular reconstruction in the treatment of microtia. Cell. Ther. Transplant. **9**, 78–82 (2020)
25. Beyer-Berjot, L., Berdah, S., Hashimoto, D.A., Darzi, A., Aggarwal, R.: A virtual reality training curriculum for laparoscopic colorectal surgery. J. Surg. Educ. **73**, 932–941 (2016)
26. Huber, T., Paschold, M., Hansen, C., Wunderling, T., Lang, H., Kneist, W.: New dimensions in surgical training: immersive virtual reality laparoscopic simulation exhilarates surgical staff. Surg. Endosc. **31**, 4472–4477 (2017)
27. Sampogna, G., Pugliese, R., Ellie, M., Vanzulli, A., Forgione, A.: Routine clinical application of virtual reality in abdominal surgery. Minim. Invasive Ther. Allied Technol. **26**, 135–143 (2017)
28. Maruthappu, M., Keogh, B.: How might 3D printing affect clinical practice. BMJ **349**, g7709 (2014)
29. Bortolotto, C., Eshka, E., Peroni, C., Orlandi, M.A., Bizzotto, N., Poggi, P.: 3D printing of CT dataset: validation of an open source and consumer-available workflow. J. Digit. Imaging **29**, 14–21 (2016)
30. Rosset, A., Spadola, L., Ratib, O.: Osirix: an open-source software for navigating in multidimensional DICOM images. J. Digit. Imaging **17**, 205–216 (2004)

# Chapter 8
# Nursing Department

**Abstract** Nurses play an important role in a hospital, whether they are nurses or nurse practitioners. They micromanage different situations, so that the doctors can focus on diagnosis, treatment plans, surgery, etc. In many hospitals, nurses are understaffed, so we need to see if AI can manage their department, so that mistakes due to fatigue, time pressure and other factors are avoided.

On May 12, 1820, the birth of an amazing woman, marked the beginning of a profound transformation in the nursing profession. We are talking about Florence Nightingale. Florence studied mathematics, language, religion, and philosophy, and all these subjects taught by her father made an impact on her future work, [1]. When she was 17, as she was strolling in a garden, she said she heard the voice of God, that told her she had an important mission. Hence, in 1844, despite her parents being against this, she decided to work in a hospital. At that time, nursing was not considered to be a decent job, [2]. After six years, she starting working as a volunteer in German and French hospitals. At 33 years old, Florence Nightingale become a hospital manager, running the Institute of sick women.

In 1854, the Crimean war begun. Britain, France, and Turkey started a war against Russia. In the battle of Alma, even if Britain defeated Russia, it had a lot of casualties. The mortality was high in British military camps. As Florence was a friend of Sydney Herbert, the minister of war, she asked him if she could take nurses and go to a British military camp. And so, she did. She took 38 nurses and went near Constantinople (Istanbul). When she arrived at those barracks, she saw that the injured soldiers were lying on the floor, as very few doctors were trying to manage the patients in a dirty environment, [3]. Making use of her statistics knowledge, she recorded the mortality rate in the hospital. She discovered that out of 1000 soldiers, 600 would die due to infectious diseases. The answer to this problem was quite simple: she needed to provide a clean environment (medical equipment, clean water, and fruits). The mortality rate decreased at first from 60 to 42%, and then to 2.2%.

Florence Nightingale is known as the Lady with the lamp, as each night she would take her lamp and go see the patients. She became famous in England, and attracted the support of Queen Victora, Prince Albert, and Prime Minister Lord



S. Belciug, *The Hospital Manager's Guide to Artificial Intelligence*, Intelligent Systems Reference Library 263, https://doi.org/10.1007/978-3-031-80314-7_8

Palmerston. At her request the Royal Institute of research on the health of military was established. She was a well-known statistician, and she invented a special pie chart. In 1860, Florence founded the first nursing school in the world, chancing the world's perspective on the nursing profession.

So, what have learned so far? That statistics is a powerful tool that can change the healthcare system into a better one. So far throughout this book we have discussed many statistical terms and tools, but we have not yet covered all the needed statistical tests. Hence, we shall continue with discussing some of the most used statistical tests and use examples from the nursing department.

***Student's t-test or the t-test***

In general, the *t*-test is used to determine whether there are statistically significant differences between two sample groups. It can be applied if the data follows a Normal distribution. If we know the null hypothesis and the test statistics follows a Student's *t* distribution, then we are using a *t*-test.

It was 1908, when William Sealy Gosset invented the *t*-test. He was a chemist and a statistician that worked at the Guinness brewery in Dublin. As one of the best graduates from Oxford, he was hired at Guinness. To make a perfect beer, Guinness needed to eliminate any element of surprise from the brewing process. Hence, Gosset started to study the quality of water, the malted barley, yeast, and hops, and the way they depended on the climate, crop variety, soil, etc. He needed to model the brewing process to be cost effective also. Hence, the only solution was to find a way to draw broad conclusions from experiments done on small samples. Recall that there was only one problem: small sample's distribution diverges from the Normal distribution.

As he started experimenting, he observed that as the number of observations grew, the smaller the error got, and the curve that represented the errors' frequencies became taller and narrower when compared to the Normal distribution, [4]. To verify whether his observations were correct or not, he had to concur with a mathematician. So, he wrote a letter to Karl Pearson. Gosset left Guiness for two semesters to learn in the Karl's Pearson Biometric Laboratory. His work was published under the pseudo name Student, since Guinness did not allow him to publish his work otherwise, [5]. The name came from a notebook he wrote on, [6].

In six months, he built the *t* table which contains the cumulative distribution function relative to *n* (sample size) and *z* (error's distribution). The higher the *z* value, the greater chances were that the test's conclusions were indeed correct. The test was popularized by Ronald Fisher.

There are three types of *t*-tests, [7]:

- *One-sample* *t*-test is used for verifying whether the mean of a sample, which is normally distributed, is equal to a value stated in the null hypothesis.
- *Paired* *t*-test is used to compare the means of a population measured at different times.
- *t*-test for *independent samples* is used for comparing the means of two independent sample groups.

**Table 8.1** Response times of a nurse to patient call bell

| Patient call | Response time |
|---|---|
| 1 | 4.5 |
| 2 | 5.2 |
| 3 | 5.1 |
| 4 | 6.0 |
| 5 | 4.8 |
| 6 | 5.3 |
| 7 | 5.5 |
| 8 | 5.4 |
| 9 | 4.9 |
| 10 | 5.0 |
| 11 | 6.1 |
| 12 | 4.7 |
| 13 | 5.2 |
| 14 | 4.6 |
| 15 | 5.3 |

Before we apply the *t*-test, we must verify whether the samples are normally distributed and have equal variances. In the case where the samples are not equal, we can apply Welch's test.

To find if two samples are similar or not, we can compute the *t*-score, which is the ratio of the difference between two sample groups and the difference within the sample groups. If the *t*-score is low then the groups are similar, otherwise they are different.

Let's present some nursing examples for you to understand how the *t*-test works. We believe that in general, nurses do not use statistics in their daily routine, but maybe they should.

*One sample t*-test

Let us suppose that the nurse supervisor wants to see if the average time it takes a nurse to respond to a patient call bell is different or not from the hospital's standard response time of 5 min, $\mu_0$. The nurse supervisor will start recording the response times (in minutes) for a sample of 15 nurse responses. The collected data is displayed in Table 8.1:

We first need to compute the sample mean, $\overline{x}$ :

$$\overline{x} = \frac{\sum_{i=1}^{n} x_i}{n} = 5.17$$

Next, we compute the sample's standard deviation, *s*. For this, we compute the differences from the mean for each data point, square them, and find the mean of the squared differences:

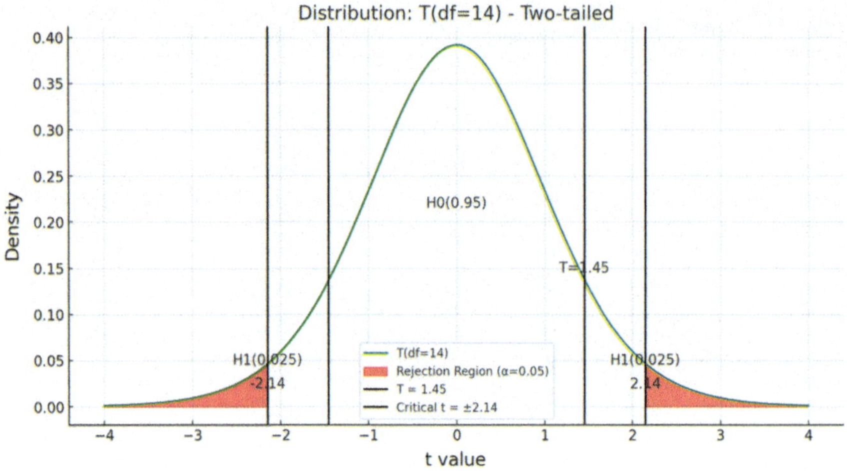


**Fig. 8.1** T distribution graph

$$s^2 = \frac{\sum_{i=1}^{n}(x_i - \overline{x})^2}{n-1} = \frac{2.9906}{14} \approx 0.2136$$

$$s = \sqrt{0.2136} \approx 0.462$$

Now, we can compute the $t$-statistic:

$$t = \frac{\overline{x} - \mu_0}{\frac{s}{\sqrt{n}}} = \frac{5.17 - 5}{\frac{0.462}{\sqrt{15}}} \approx 1.424$$

Having the statistic value 1.424, we check Table A11 from the Appendix for 14 degrees of freedom at 0.05. The critical value for this significance level (two-tailed) is 2.145, therefore because 1.424 is less than 2.145, we accept the null hypothesis (Fig. 8.1). This means, that the average response of the nurse is not significantly different from the hospital's standard.

*Paired t*-test

Let's presume that in our hospital we are implementing a new training program for the nursing staff. Our aim is to reduce the time it takes for a nurse to administer a certain medication. To see whether the training program is effective or not, we measure the time it takes for each nurse to administer the treatment before and after they undergo the training program. We collect the data on 10 nurses (Table 8.2). The time in measured in minutes.

First, we compute the mean, $\mu_d$, and standard deviation of the differences between the time before and after the training.

**Table 8.2** Time to administer medication before and after training

| Nurse | Score before training | Score after training |
|---|---|---|
| 1 | 12 | 10 |
| 2 | 15 | 13 |
| 3 | 14 | 12 |
| 4 | 13 | 11 |
| 5 | 16 | 14 |
| 6 | 14 | 13 |
| 7 | 15 | 12 |
| 8 | 16 | 14 |
| 9 | 17 | 15 |
| 10 | 18 | 16 |

$$\mu_d = \frac{\sum d_i}{n} = \frac{2+2+2+2+2+1+3+2+2+2}{10} = \frac{20}{10} = 2$$

The standard deviation of the differences is:

$$s_d = \sqrt{\frac{\sum (d_i - \overline{d})^2}{n-1}} = \sqrt{\frac{2}{9}} \approx 0.471.$$

Let us state the two hypotheses:

$H_0$: The training program has no effect on the time it takes for nurses to administer the treatment, hence $\mu_d = 0$.

$H_1$: The training program has a significant effect, hence $\mu_d \neq 0$.

To compute the test statistic for the paired *t*-test we use the following formula:

$$t = \frac{\mu_d}{s_d/\sqrt{n}} = \frac{2}{0.471/\sqrt{10}} \approx 13.42.$$

Using the significance level 0.05 and 9 degrees of freedom, we look to the critical *t*-value for the two-tailed test, Table A12 in the Appendix, and see the value 2.262. Since $|13.42| > 2.262$, we reject the null hypothesis (Fig. 8.2). We now have sufficient evidence to conclude that the training program indeed reduces the time it takes a nurse to administer the medication.

***t*-test for independent samples**

Let us suppose that we want to evaluate whether there is a difference in the average number of patient rounds completed during a night shift and those completed during a day shift. We select 10 nurses randomly from each shift and record the number of patient rounds they complete during a shift (Table 8.3).

Our working hypotheses are:

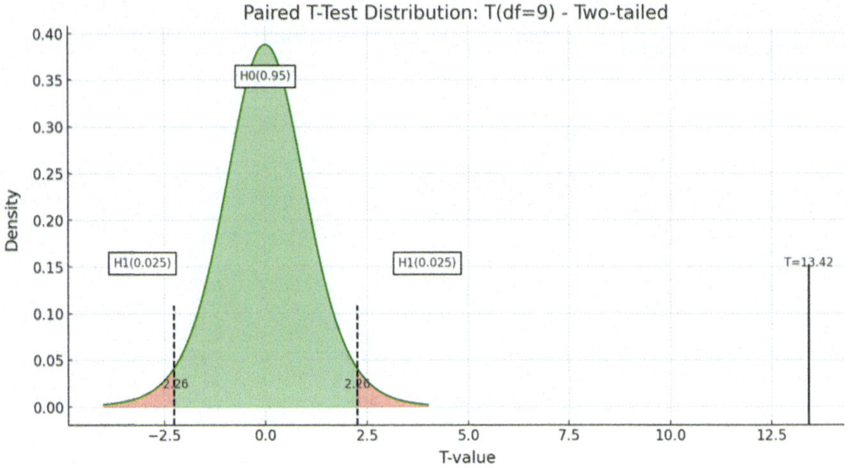


**Fig. 8.2** Paired $t$-test distribution

**Table 8.3** Number of patient rounds completed during day and night shifts

| Nurse | Day shift | Night shift |
|---|---|---|
| 1 | 20 | 18 |
| 2 | 22 | 19 |
| 3 | 23 | 17 |
| 4 | 21 | 16 |
| 5 | 25 | 20 |
| 6 | 24 | 21 |
| 7 | 23 | 19 |
| 8 | 22 | 18 |
| 9 | 21 | 17 |
| 10 | 20 | 16 |

$H_0$ : The average number of patient rounds completed per day shifts is equal to the number of patient rounds completed per night shift ($\mu_{day} = \mu_{night}$).

$H_1$: The average number of patient rounds completed per day shift is not equal to the number of patients completed per night shift ($\mu_{day} \neq \mu_{night}$).

The mean of the day shift is:

$$\mu_{day} = \frac{20 + 22 + 23 + 21 + 25 + 24 + 23 + 22 + 21 + 20}{10} = 22.1\ rounds$$

The mean of the night shift is:

$$\mu_{night} = \frac{18 + 19 + 17 + 16 + 20 + 21 + 19 + 18 + 17 + 16}{10} = 18.1 \textit{ rounds}$$

The variance for the day shift is:

$$s_{day} = \frac{\sum \left(X_{day_i} - \mu_{day}\right)^2}{n_{day} - 1} = 2.77.$$

The variance for the night shift is:

$$s_{night} = \frac{\sum \left(X_{night_i} - \mu_{night}\right)^2}{n_{night} - 1} = 2.77.$$

We can see that both groups have equal variances.

Recall, that before we apply the $t$-test for independent samples we need to verify the normality of the data and the equality of variances. We have already demonstrated that both samples have equal variances, hence, we shall apply a normality test, for instance, the Shapiro–Wilk test and see whether the samples are governed by the Normal distribution or not. For both groups the corresponding $p$-level is 0.692, indicating that we cannot reject the normality assumption.

The $t$ statistic for an independent samples $t$-test is computed as:

$$t = \frac{\mu_{day} - \mu_{night}}{\sqrt{\frac{s^2_{day}}{n_{day}} + \frac{s^2_{night}}{n_{night}}}} = \frac{22.1 - 18.1}{\sqrt{\frac{2.77}{10} + \frac{2.77}{10}}} = \frac{4}{\sqrt{0.277 + 0.277}}$$

$$= \frac{4}{\sqrt{0.554}} \approx 5.38.$$

Using Table A12 from the Appendix, we search for the test statistic 5.38 and 18 degrees of freedom. We see that the $p$-level is less than 0.05, hence we reject the null hypothesis and accept the alternative one, which means that indeed there are statistical differences between the day and night shift nurses.

In Fig. 8.3 we present the plot for this example. We can see the $t$-distribution with 18 degrees of freedom represented by the green curve. The rejection regions for a two-tailed test with 95% confidence level are depicted with a red shade, whereas the dashed black lines represent the critical values. You can see the computed $t$ statistics represented by the gray line which falls outside the critical region, demonstrating the fact that we should reject the null hypothesis.

***Sign test***

If our data is not governed by the Normal distribution, we cannot proceed with applying the $t$-test, thus we need to find an alternative. One possible alternative is a non-parametric or a free distribution test, such as the sign test. Having a hypothesized mean, we compare that mean with the sample mean, and if they are approximately

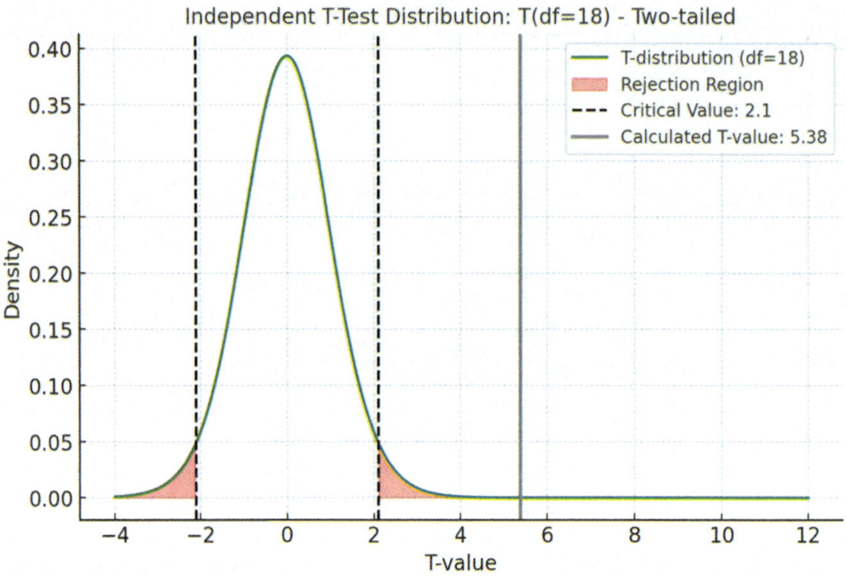


**Fig. 8.3** *T*-test for independent sample plot (day shift vs. night shift)

equal, then we have the same number of instances above and below the mean. We assess the likelihood of observing frequencies above and below the sample mean, given that the true probability of exceeding the expected intake is $p = \frac{1}{2}$, by applying the Binomial distribution.

Let's see how we can apply the sign test on a practical example. Let us suppose that in our hospital the nursing department wants to test whether a new medication administration procedure does indeed reduce the time it takes a nurse to deliver the treatment. The chief nurse collects the data regarding the time to administer medication of 12 nurses before and after implementing the new procedure.

Table 8.4 presents the time (in minutes) it takes each nurse to administer the treatment before and after the procedure was implemented. The table contains the difference before and after, and the sign.

We are interested in determining whether the median time to administer the medications has decreased or not. The working hypotheses are:

$H_0$ : The median difference between the times before and after the procedure is zero (there is no difference whatsoever).

$H_1$ : The median time to administer the medication after the procedure has been implemented has decreased.

The first step we need to undertake is to ignore the zero differences. In our case, nurses 3, 4, 11, and 12 have no change in their performance, thus they will be excluded from the analysis. We have 8 more observations left in our study. We will count the positive and negative signs:

**Table 8.4** Recorded time before/after the procedure was implemented

| Nurse | Before procedure | After procedure | Difference (before-after) | Sign (±) |
|---|---|---|---|---|
| 1 | 25 | 20 | +5 | + |
| 2 | 30 | 28 | +2 | + |
| 3 | 22 | 22 | 0 | 0 |
| 4 | 35 | 35 | 0 | 0 |
| 5 | 28 | 29 | −1 | − |
| 6 | 40 | 34 | +6 | + |
| 7 | 32 | 30 | +2 | + |
| 8 | 24 | 22 | +2 | + |
| 9 | 29 | 27 | +2 | + |
| 10 | 31 | 29 | +2 | + |
| 11 | 27 | 27 | 0 | 0 |
| 12 | 26 | 26 | 0 | 0 |

- Positive signs (+): 7 nurses took less time to administer the medicine after the procedure was implemented.
- Negative signs (−): 1 nurse took more time to administer the medicine after the procedure was implemented.

As stated above, we need to make use of the Binomial distribution to compute the probability of observing more positive signs under the null hypothesis. We assume that each sign is equally likely, $p = 0.5$. We have:

- Number of trials ($n$): 8 (recall that we have eliminated the non-zero differences).
- Number of successes ($k$): 7 (the number of positive signs).
- Probability of success under the null hypothesis ($p$): 0.5.

The Binomial test formula needed to compute the probability of obtaining $k$ positive signs is:

$$P(k) = \binom{n}{k} p^k (1-p)^{n-k}.$$

We need to compute the probability of obtaining 7 or more positive signs, thus we will sum up the probabilities for 7 and 8 signs.

For $k = 7$, we have:

$$P(7) = \binom{8}{7} (0.5)^7 (0.5)^1 = \frac{8!}{7!(8-7)!} \times (0.5)^8 = 0.03125.$$

For $k = 8$, we have:

$$P(8) = \binom{8}{8}(0.5)^8 = 1 \times (0.5)^8 = 0.0039.$$

The sum of probabilities for $k \geq 7$ is:

$$P(k \geq 7) = P(7) + P(8) = 0.03515.$$

The $p$-value for observing 7 or more positive signs is 0.03515, (Table A11 from the Appendix). In conclusion, we can reject the null hypothesis. This means that the new procedure has significantly reduced the time of administering treatments.

Another alternative to the $t$-test is the Wilcoxon signed rank sum test.

***Wilcoxon signed rank sum test***

The Wilcoxon signed rank sum test takes into consideration the magnitude of the differences between two sample data, not just whether the value of an observation is greater or less than a hypothesized mean. The test is conducted by applying the following steps:

1. Calculate the difference between the known mean and every observation.
2. Rank the absolute values, ignoring the signs.
3. Compute the sum of the ranks corresponding to the differences that are above (or below) the hypothesized mean.

If our samples' sizes are below 25, we use Table A14 from the Appendix to find whether we accept or reject the null hypothesis. If the size is above 25, the Normal distribution governs our data, with mean:

$$\mu = \frac{n(n+1)}{4},$$

and standard deviation:

$$\sigma = \sqrt{\frac{n(n+1)(2n+1)}{24}}.$$

In this case, the statistic value cand be found in Table A14, from the Appendix.

Let's consider the following example: our hospital has implemented a new patient care protocol, and we wish to acknowledge if the satisfaction scores are improved or not. Before implementing the new protocol, the satisfaction scores from 10 randomly selected patients were collected. The scores range on a scale from 1 to 10 (where 10 is the highest level of satisfaction). Table 8.5 presents the collected data and the difference between before and after.

After computing the differences, we need to rank the absolute difference, while ignoring the sign. The ranked differences are presented in Table 8.6.

The ranking procedure goes like this. We have the smallest absolute difference of 1, which appears 5 times for the following patients 4, 7, 8, 9 and 10. The absolute

**Table 8.5** Patient satisfaction scores before and after the new protocol has been implemented

| Patient | Satisfaction before | Satisfaction after | Difference (before-after) |
|---|---|---|---|
| 1 | 6 | 8 | − 2 |
| 2 | 7 | 9 | − 2 |
| 3 | 8 | 10 | − 2 |
| 4 | 5 | 6 | − 1 |
| 5 | 7 | 7 | 0 |
| 6 | 6 | 8 | − 2 |
| 7 | 8 | 9 | − 1 |
| 8 | 7 | 8 | − 1 |
| 9 | 9 | 10 | − 1 |
| 10 | 6 | 7 | − 1 |

**Table 8.6** Ranked differences

| Patient | Difference | Absolute Difference | Rank | W- | W + |
|---|---|---|---|---|---|
| 1 | − 2 | 2 | 7.5 | | 7.5 |
| 2 | − 2 | 2 | 7.5 | | 7.5 |
| 3 | − 2 | 2 | 7.5 | | 7.5 |
| 4 | − 1 | 1 | 3 | | 3 |
| 5 | 0 | 0 | – | | |
| 6 | − 2 | 2 | 7.5 | | 7.5 |
| 7 | − 1 | 1 | 3 | | 3 |
| 8 | − 1 | 1 | 3 | | 3 |
| 9 | − 1 | 1 | 3 | | 3 |
| 10 | − 1 | 1 | 3 | | 3 |
| | | | | 0 | 45 |

difference of 2 appears in 4 times in the following patients 1, 2, 3 and 6. The rank for patients that have 1 as difference is computed as:

$$\frac{1+2+3+4+5}{5} = 3,$$

while the rank for patients that have 2 as difference is computed as:

$$\frac{6+7+8+9}{4} = 7.5.$$

Since all the differences are negative, we sum up the ranks for the negative differences:

**Table 8.7** Patient satisfaction scores (ward A and B)

| Ward A satisfaction scores | Ward B satisfaction scores |
|---|---|
| 8 | 6 |
| 6 | 5 |
| 7 | 7 |
| 9 | 8 |
| 5 | 4 |
| 7 | 6 |

$$W = 7.5 + 7.5 + 7.5 + 3 + 7.5 + 3 + 3 + 3 + 3 = 45.$$

The sum of signed ranks is 45.
The formula for the mean of Wilcoxon signed-rank test distribution is:

$$\mu_W = \frac{n(n+1)}{4} = \frac{9(9+1)}{4} = 22.5$$

The standard deviation of $W$ is:

$$\sigma_W = \sqrt{\frac{n(n+1)(2n+1)}{24}} = 8.215$$

All that is left for us to do is to compute the $z$-statistic using the formula:

$$z = \frac{W - \mu_W}{\sigma_W} = 2.67.$$

Using Table A14 from the Appendix we see that the associated $p$-value is 0.0074, which is less than 0.05, which means that we reject the null hypothesis.

**Mann–Whitney U Test**

If we wish to compare two independent samples, but the data does not fulfil the normality assumption, we can use a non-parametric alternative for the $t$ test, the Man-Whitney-Wilcoxon test or simply put the Mann–Whitney U test. The Mann–Whitney U test ranks all the instances from the two samples as it is only one sample. After this step is over, the ranks are summed up and the corresponding $p$-level is searched in the corresponding table (Table A15 from the Appendix).

Let's imagine the following scenario. In our hospital we want to compare the patient satisfaction levels between two different wards (ward A and ward B). In both wards we have implemented the new patient care protocol discussed in the above example. The satisfaction level was recorded from randomly selected patients in both wards. The satisfaction level is measured on a scale from 1 to 10, 10 being the highest satisfaction. Table 8.7 presents the recorded scores.

**Table 8.8** Ranked observations of patients' scores

| Ward | Satisfaction score | Rank |
|---|---|---|
| A | 8 | 9 |
| A | 6 | 4.5 |
| A | 7 | 7 |
| A | 9 | 11 |
| A | 5 | 2.5 |
| A | 7 | 7 |
| B | 6 | 4.5 |
| B | 5 | 2.5 |
| B | 7 | 7 |
| B | 8 | 9 |
| B | 4 | 1 |
| B | 6 | 4.5 |

The working hypotheses are:

$H_0$ : the median patient satisfaction score in ward A is equal to the score in ward B.

$H_1$ : the median patient satisfaction score in ward A is not equal to the score in ward B.

In Table 8.8 we have combined and ranked all the observations in ascending order. The sum of ranks for ward A is:

$$R_A = 9 + 4.5 + 7 + 11 + 2.5 + 7 = 41$$

and for ward B is:

$$R_B = 4.5 + 2.5 + 7 + 9 + 1 + 4.5 = 28.5$$

Next, we compute the U statistics for each group:

$$U_A = n_A n_B + \frac{n_A(n_A + 1)}{2} - R_A = 16$$

$$U_B = n_A n_B + \frac{n_B(n_B + 1)}{2} - R_B = 28.5.$$

where $n_A$ is the number of observations in ward A, and $n_B$ is the number of observations in ward B.

The Mann–Whitney U statistic takes the minimum between the two U values, hence:

$$U = \min(U_A, U_B) = \min(16, 28.5) = 16.$$

The mean and standard deviation for the distribution of U are:

$$\mu_U = \frac{n_A n_B}{2} = \frac{6 \times 6}{2} = 18.$$

$$\sigma_U = \sqrt{\frac{n_A n_B (n_A + n_B + 1)}{12}} = 6.24$$

The $z$-statistics for the Mann–Whitney U test is:

$$z = \frac{U - \mu_U}{\sigma_U} = -0.32.$$

The corresponding $p$-level is 0.75 (two-tailed test), thus being greater than 0.05, we accept the null hypothesis. There is no statistically significant difference in patient satisfaction scores between ward A and ward B.

**Welch test for unequal variances or simply put the Welch *t*-test**

If our samples do not have equal variances, we can use the Welch $t$-test as an alternative to the $t$-test. The Welch $t$-test differs from the $t$ test in what regard the degrees of freedom. Please note that both tests presume that the samples are governed by the Normal distribution. If the samples have equal variances, we expect the same result for the Welch test as well as for the $t$-test. Otherwise, the results should be different. The Welch's $t$ statistic is computed as:

$$t = \frac{\mu_1 - \mu_2}{\sqrt{se_1^2 + se_2^2}.}$$

To approximate the degrees of freedom $d$ we use the Satterthwaite-Welch equation, [8–10]:

$$d \approx \frac{\left(\frac{s_1^2}{n_1} + \frac{s_2^2}{n_2}\right)^2}{\frac{\left(\frac{s_1^2}{n_1}\right)^2}{n_1 - 1} + \frac{\left(\frac{s_2^2}{n_2}\right)^2}{n_2 - 1}}.$$

Let's imagine the following scenario. A hospital wants to determine if there is a significant difference between the number of hours nurses work per week in the Emergency Room (ER) versus the number of hours they work in the Pediatrics Department (PD). Since the nature of their work is different, we expect that the variances are not equal between the two groups. Table 8.9 presents the data collected in terms of hours worked per week from both departments.

**Table 8.9** Hours worked per week in the ER and PD departments

| ER | PD |
|---|---|
| 40 | 35 |
| 42 | 37 |
| 39 | 38 |
| 45 | 36 |
| 41 | 34 |
| 44 | 32 |
| 46 | 33 |
| | 36 |

Our working hypotheses are:

$H_0$ : the mean number of hours worked per week is the same in both departments ($\mu_{ER} = \mu_{PD}$).

$H_1$ : the mean number of hours worked per week is not the same in both departments ($\mu_{ER} \neq \mu_{PD}$).

Below, we present the statistics for each department:

$$\mu_{ER} = 42.43$$

$$s_{ER}^2 = 6.95$$

$$\mu_{PD} = 35.13$$

$$s_{ER}^2 = 4.13$$

Using the formulas presented above, we compute the $t$ statistics = 5.95, and the degrees of freedom 11.25. The associated $p$-level is 0.000008 less than 0.05 (Table A16 from the Appendix), thus we reject the null hypothesis. There is a statistically significant difference in the average numbers of hours worked by the nurses from the Emergency Room and Pediatrics Department.

Until now, we have covered tests that compare two groups of observations. But what happens if we have multiple groups of observations, and we want to compare them? We cannot keep applying the $t$-test on two-by-two groups. Our proposed solutions are the *one-way analysis of variance*, also known as *one way ANOVA* or the *two-way ANOVA*. The difference between the two methods is given by the number of independent variables used. The one-way ANOVA uses one independent variable that has two levels or groups, whereas the two-way ANOVA uses two independent variables that have multiple levels or groups.

One-way ANOVA compares the mean of multiple data samples using the $F$-distribution. The null hypothesis states that there aren't significant differences between the means. Note that the one-way ANOVA just gives insights regarding the existence of differences among the groups but does not pinpoint where the differences are. Thus, after applying ANOVA you should always follow-up with a post-hoc test. More on post-hoc tests later. An important point that you need to remember is that, when you are conducting, this type of test, you are dividing the total variability of the sample data into different components, each one corresponding to a distinct source of variation.

One-way ANOVA uses the sum of the observations in each group sample. If we have $q$ groups, and in each group, we have $n_i$ observations, then we compute:

$\mu_i$—the means of the $i$th sample group;

$S_i$—the sum of squares of observations from the $i$th sample group;

$$T = \sum_{i=1}^{q} n_i \cdot \mu_i;$$

$$S = \sum_{i=1}^{q} S_i;$$

$$N = \sum_{i=1}^{q} n_i.$$

The ANOVA's sum of squares is computed as:

$$B = \sum_{i=1}^{q} n_i \mu_i^2 - \frac{T^2}{N}.$$

The sum of squares within each group is computed as:

$$W = S - \sum_{i=1}^{q} n_i \mu_i^2.$$

The total sum of squares is:

$$B + W = S - \frac{T^2}{N}.$$

The degrees of freedom within the groups are $n - q$, and the degrees of freedom between the groups is $k - 1$. To calculate the mean squares, we need to determine the ratio of the sum of squares to the degrees of freedom. As for the square root within the groups, it corresponds to the residual standard deviation.

In the case we do not have access to the raw data, we can still use ANOVA if we have the means and the standard deviations. We can compute $B$ and $T$ as we have

**Table 8.10** Recovery times

| Surgery ward | Cardiology ward | Orthopedic ward |
|---|---|---|
| 8 | 12 | 10 |
| 10 | 14 | 8 |
| 7 | 13 | 9 |
| 9 | 15 | 11 |
| 6 | 14 | 9 |

shown above. $W$ is computed as:

$$W = \sum_{i=1}^{q} (n_i - 1)s_i^2.$$

Let us presume that we want to evaluate whether there is a significant difference in the average patient recovery times measured in days, across three wards: Surgery Ward, Cardiology Ward, and Orthopedic Ward. Table 8.10 presents the recovery time for patients that have been selected randomly from each ward:

Our working hypotheses are:

$H_0$ : The mean recovery time in days is the same across all wards ($\mu_{surgery} = \mu_{cardiology} = \mu_{orthopedic}$).

$H_1$ : The mean recovery time in days is not the same across all wards ($\mu_{surgery} \neq \mu_{cardiology} \neq \mu_{orthopedic}$).

The group means are:

$$\mu_{surgery} = \frac{8 + 10 + 7 + 9 + 6}{5} = 8.$$

$$\mu_{cardiology} = \frac{12 + 14 + 13 + 15 + 14}{5} = 13.6.$$

$$\mu_{orthopedic} = \frac{10 + 8 + 8 + 11 + 9}{5} = 9.4.$$

The overall mean is computed by combining all data points from all wards:

$$\mu_{overall} = \frac{8 + 10 + 7 + 9 + 6 + 12 + 14 + 13 + 15 + 15 + 14 + 10 + 8 + 9 + 11 + 9}{15}$$
$$= 10.33.$$

The sum of squares between groups (SSB) is:

$$SSB = n_{surgery}\left(\mu_{surgery} - \mu_{overall}\right)^2 + n_{cardiology}\left(\mu_{cardiology} - \mu_{overall}\right)^2 + n_{orthopedic}\left(\mu_{orthopedic} - \mu_{overall}\right)^2 = 84.93$$

The sum of squares withing groups (SSW) is:

$$SSW = \sum_{i=1}^{n_{surgery}}\left(x_{surgery} - \mu_{surgery}\right)^2 + \sum_{i=1}^{n_{cardiology}}\left(x_{cardiology} - \mu_{cardilogy}\right)^2 + \sum_{i=1}^{n_{orthopedic}}\left(x_{orthopedic} - \mu_{orthopedic}\right)^2 = 20.40.$$

The mean square between groups (MSB) is:

$$MSB = \frac{SSB}{k-1} = 42.47$$

The means square within groups (MSW) is:

$$MSW = \frac{SSW}{n-k} = 1.70.$$

The $F$-statistics is computed by dividing the mean square between groups by the means square within groups:

$$F = \frac{MSB}{MSW} = 24.98$$

The critical $F$-value is 3.89. The associated $p$-level is 0.00053, which indicates that there is a statistically significant difference in the average recovery times among these three wards. We reject the null hypothesis. Figure 8.4 presents a power F-distribution graph. The gray area represents the region where we accept the null hypothesis, the blue area represents the rejection region for $\alpha = 0.05$, whereas the red area represents the Type II error, where we fail to reject the null hypothesis when it is false. The power of the test is indicated by the green area and indicated the probability of correctly rejecting the null hypothesis. The dashed line represents the critical $F$-value.

Let's see how a two-way ANOVA example works. We want to investigate the impact of the shift type (day vs. night), the nurses' experience level (junior, mid-level, senior), and time to administer the treatment. We collect the time it takes to administer the treatment depending on the shift type and experience level. Table 8.11 depicts the data.

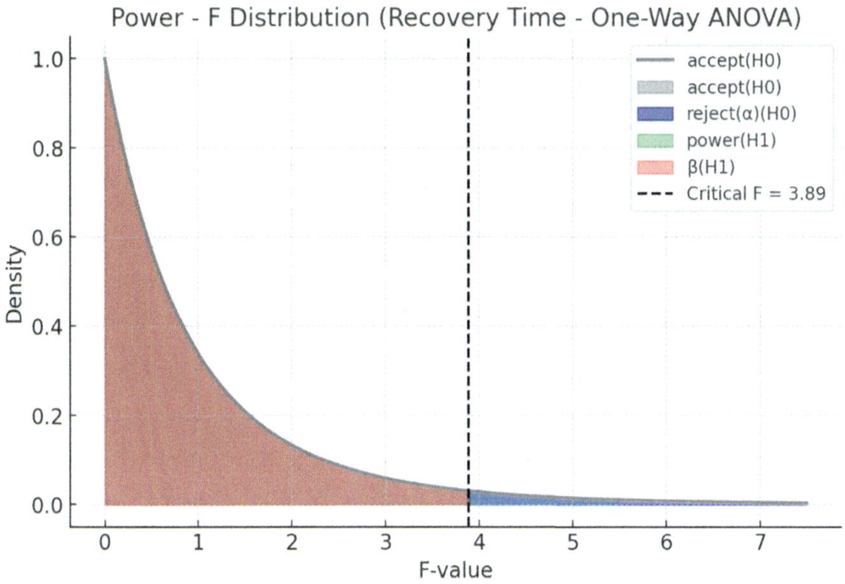


**Fig. 8.4** Power F-distribution graph of recovering times

**Table 8.11** Time to administer treatment depending on the shift type and experience level

| Experience level | Day shift | Night SHIFT |
|---|---|---|
| Junior | 12, 15 | 20, 22 |
| Mid-level | 10, 13 | 18, 21 |
| Senior | 8, 9 | 16, 17 |

We wish to determine whether there are significant differences between the time taken to administer the treatment based on the shift type, based on the experience level, and whether there is an interaction effect between the shift type and experience level.

Let us state the working hypotheses:

1. Effect of shift type (day vs. night):

$H_0$ : the mean times for the day and night shifts are equal.

$H_1$ : the mean times for the day and night shifts are not equal.

2. Effect of experience level (junior, mid-level, senior):

$H_0$ : the mean times for the three experience levels are equal.

$H_1$ : the mean times for the three experience levels are not equal.

3. Interaction effect between shift type and experience level:

**Table 8.12** Results of two-way ANOVA

| Source of variation | Sum of squares | Degrees of freedom | F-statistic | *p*-value |
|---|---|---|---|---|
| Shift (day vs. night) | 184.08 | 1 | 66.94 | 0.00018 |
| Experience level | 46.17 | 2 | 8.39 | 0.01825 |
| Interaction (shift—experience) | 0.17 | 2 | 0.03 | 0.9703 |
| Residual | 16.5 | 6 | | |

$H_0$ : there is no interaction effect between shift type and experience level.

$H_1$ : there is significant interaction effect between the shift type and experience level.

The two-way ANOVA results are presented in Table 8.12:

We can see from Table 8.12, that the *p*-value is 0.00018 for the shift type, value which is much smaller than 0.05. Thus, we will reject the null hypothesis, because there is significant evidence that there is a difference in the time taken to administer medications based on the shift (day vs. night). The *p*-value for the experience level is 0.01825, which is also smaller than 0.05, hence we once again reject the null hypothesis. There is a significant difference in the time taken to administer the treatment based on the level of experience the nurse has. For the interaction between the shift type and experience, we see that the *p*-value is 0.97, which is larger than 0.05. Hence, we accept the null hypothesis, that indicated that there is no significant interaction effect between the shift type and experience level when it comes to the time taken to administer the treatment. Figure 8.5 presents a similar plot as in the case of one-way ANOVA.

***Post-hoc tests***

After apply ANOVA, you should apply a post-hoc test. A post-hoc test shows exactly where are the differences between the tested groups. A post-hoc test is able to control the error rate. The error rate can be family-wise or experiment-wise. Recall the type I error rate, if we are performing one test the error rate is equal to the chosen significance level, but if we are performing multiple tests, then we increase the chance of obtaining a false positive. ANOVA performs multiple tests, so our chances in this case can increase. If we have 4 groups, we make 6 comparisons. If we are performing one test we have an experiment-wise error, if we are performing multiple comparisons we have a family-wise error. In what follows, we shall present the most commonly used post-hoc tests.

*Bonferroni correction*

We apply the Bonferroni correction when we are conducting multiple independent or dependent statistical tests simultaneously. As we mentioned earlier, when running multiple tests, the likelihood of us obtaining a false positive result is higher with each test. To counter this, the Bonferroni correction adjusts the significance threshold to $\frac{\alpha}{n}$, where $n$ is the number of tests. However, there is a drawback. As the test increases the risk of Type II errors (false negatives), it reduces the statistical power. The Bonferroni

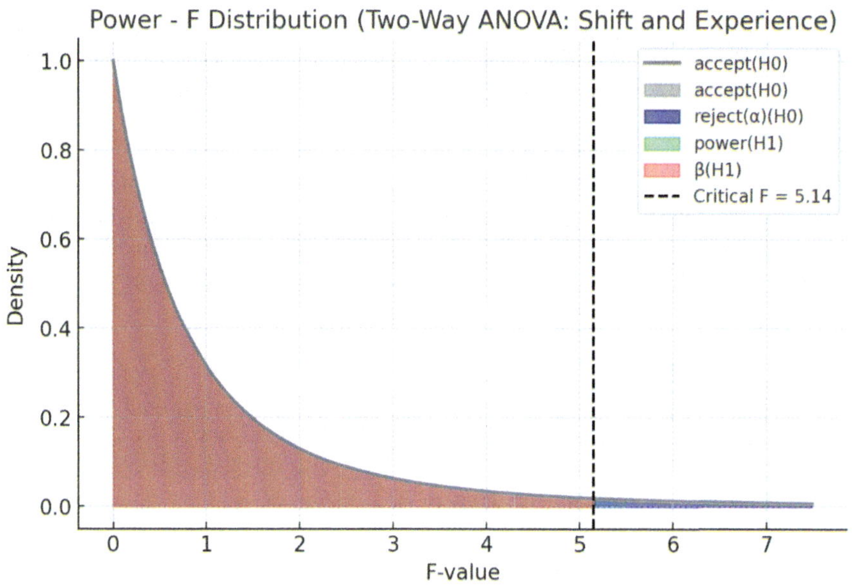


**Fig. 8.5** Power $F$-distribution plot for two-way ANOVA example

correction overcompensates the Type I errors, [11, 12]. Due to the drawback, this test is often considered to be overly conservative. Thus, we indicate its flexible alternative Holm's Sequential Bonferroni method, which is less strict and offers a higher statistical power, [13].

*Fisher's least significant difference*

In 1935, Fisher proposed a new method to be used when the null hypothesis is rejected: the least significant difference. The method calculates the minimum difference between two means that must be exceeded for the result to be considered significant. Using this, we can make a direct comparison between the means of different groups. Any difference that is larger than the computed threshold is deemed statistically significant. Further details can be found in [14–16].

*Duncan's new multiple range test*

Duncan's test identifies pair of means that differ by using the $Q$-value and not the $t$-value. When we compare a large number of mean pairs, it is preferred to use Duncan, and not Fisher (even if they are quite similar), since it has been proven to be more effective. The test was developed as a more powerful alternative to the Student–Newman–Keuls test. For further details we refer the reader to [17, 18].

*Student–Newman–Keuls*

This method is a stepwise multiple comparison post-hoc test that uses studentized range distribution. Even if this method can detect significant differences between means, it has the down flaw to incorrectly reject the null hypothesis when it is actually true. One should be cautious when using it, since it has a lower power than ANOVA, which means that ANOVA might identify differences, but this test not, [19–21].

*Scheffe*

This test is typically used when we are dealing with unequal sample sizes. The difference between this test and other post-hoc tests, is that it does not rely on pairwise comparisons, and it can be applied on both simple and complex mean comparisons. The test is flexible, but it has low power, so use it with caution. For more information see [22, 23].

*Tukey's honest significant difference (HSD)*

This type of test uses the studentized range distribution to compare all the possible pair of means. While it evaluates the pairwise comparisons, all at the same time, it is able to identify group mean differences that exceed the expected standard error. Tukey HSD can be used after ANOVA or as a standalone method. This is possible due to its ability to control the Type I error. Further details regarding this test can be found in [24, 25].

*Dunnett's correction*

When using Dunnett's correction, a control groups is required. This test compares the means of the experimental groups specifically against the control groups' mean to determine whether there are any statistically significant differences. For further information regarding this test, we refer the reader to [26, 27].

*Benjamini–Hochberg test*

This test is useful when conducting post-hoc tests on a large number of comparisons, because the chance of obtaining a significant result increases as the number of tests performed increases also. The test controls the false discovery rate. For more details on this topic, we refer the readers to [28, 29].

***Kruskal–Wallis Test***

If the data does not fulfill the assumption needed to perform ANOVA, you can use the alternative Kruskal–Wallis test, [30–33]. This test uses the ranks of the values, and not the values themselves. The Kruskal–Wallis test compares the medians of the groups, and not the means. We have the following working hypotheses:

$H_0$ : the groups have equal medians.

$H_1$ : the groups do not have equal medians.

Let us explain the theory through an example. We want to compare whether there are significant differences between the patient satisfaction scores across three different wards: surgery, cardiology, and orthopedy. We collect the scores from different patients. The scores range from 1 to 10, where the higher scores indicate greater satisfaction. Table 8.13 presents the collected data.

**Table 8.13** Patient satisfaction scores

| Surgery | Cardiology | Orthopedy |
|---|---|---|
| 7 | 9 | 6 |
| 8 | 8 | 7 |
| 6 | 10 | 8 |
| 9 | 9 | 7 |
| 7 | 10 | 6 |

**Table 8.14** Patients' scores and associated ranks

| Score | Rank |
|---|---|
| 6 | 1 |
| 6 | 1 |
| 6 | 1 |
| 6 | 1 |
| 7 | 5.5 |
| 7 | 5.5 |
| 7 | 5.5 |
| 7 | 5.5 |
| 8 | 9 |
| 8 | 9 |
| 8 | 9 |
| 9 | 12.5 |
| 9 | 12.5 |
| 9 | 12.5 |
| 10 | 14.5 |
| 10 | 14.5 |

First, we need to combine the satisfaction scores from the three wards and assign a rank to each score. Table 8.14 presents the scores and their associated ranks.

We next need to compute the rank sum of each ward by summing up the ranks corresponding to each group.

$$R_{surgery} = 5.5 + 9 + 1 + 12.5 + 5.5 = 33.5$$

$$R_{cardiology} = 12.5 + 9 + 14.5 + 12.5 + 14.5 = 63$$

$$R_{orthopedy} = 1 + 5.5 + 9 + 5.5 + 1 = 22$$

We compute the $H$-statistic for the Kruskal Wallis test as:

$$H = \frac{12}{N(N+1)} \sum \left( \frac{R_i^2}{n_i} \right) - 3(N+1)$$

where $N$ is the total number of observations, $R_i$ is the rank sum for group $I$, and $n_i$ is the number of observations in group $i$. Thus, the $H$-statistic will be:

$$H = \frac{12}{15(15+1)} \left( \frac{33.5^2}{5} + \frac{63^2}{5} + \frac{22^2}{5} \right) - 3(15+1) = 7.75.$$

Using the $\chi^2$ distribution table (Table A10 from the Appendix) for 2 degrees of freedom, we find that the corresponding $p$-value is 0.0175. We reject the null hypothesis, because there are statistically significant differences in patient satisfaction scores between the three wards.

This ends the chapter regarding how we can optimize the nursing department. Besides the presented tools, hospitals can acquire robot nurses that are used to administer treatment, etc. Even if they are less prone to mistakes, sometimes, the warm touch of a human hand does more good than any medicine.

## References

1. Alligood, M.R., Tomey, A.M.: Nursing Theorists and Their Work, 6th edn. Mosby, Singapore (2006)
2. McEven, M., Wills, E.M.: Theoretical Basis for Nursing. Lippincott, Philadephia (2011)
3. Asad Zandi, M.: Concepts and Theories of Nursing with a Sound of Heart Model. Heidari Publication, Tehran (2014)
4. Zabell, S.L., Stephen, S.M., Aldrich, J., Edwards, A.W.F., Seneta, E., Diaconis, P., Lehmann, E.: On student's 1908 article: the probable error of a mean. J. Am. Stat. Assoc. **103**(481), 1–20 (2008)
5. Student: The probable error of a mean. Biometrika **6** (1), 1–25 (1908)
6. Ziliak, S.T.: Retrospectives: guinnessometrics: the economic foundation of "student's *t*". J. Econ. Persp. **22**(4), 199–216 (2008)
7. Altman, D.G.: Practical Statistics for Medical Research. Chapman and Hall, London (1991)
8. Welch, B.L.: The significance of the difference between two means when the population variances are unequal. Biometrika **39**, 350–362 (1938)
9. Satterthwaite, F.E.: An approximate distribution of estimates of variances components. Biometr. Bull. **2**(6), 110–114 (1946)
10. Welch, B.L.: The generalization of Student's problem when several different population variances are involved. Biometrika **34**, 28–35 (1947)
11. Cook, T.: Introduction to Statistical Methods for Clinical Trials, 1st edn. Chapman and Hall, CRC, London (2005)
12. Wheelan, C.: Naked Statistics. W.W. Norton & Company, New York (2014)
13. Holm, S.: A simple sequential rejective multiple test procedure. Scand. J. Stat. **6**, 65–70 (1979)
14. Dodge, Y.: The Concise Encyclopedia of Statistics. Springer, Berlin (2008)
15. Gonick, L.: The Cartoon Guide to Statistics. Harper Perennial, New York (1993)
16. Klein, G.: The Cartoon Introduction to Statistics. Hill & Wamg, New York (2013)
17. Duncan, D.B.: Multiple range and multiple F tests. Biometrics **11**(1), 1–42 (1995). https://doi.org/10.2307/3001478

18. Shaffer, J.P.: A semi-Bayesian study of Duncan's Bayesian multiple comparison procedure. J. Stat. Pann. Infer. **82**(1–2), 197–213 (1999). https://doi.org/10.1016/S0378-3758(99)00042-7
19. Student: Errors of routine analysis. Biometrika **19** (1–2), 151–164 (1927). https://doi.org/10.2307/2332181
20. Newman, D.: The distribution range in samples from a normal population expressed in terms of independent estimate of standard deviation. Biometrika **31** (1), 20–30 (1939). https://doi.org/10.10193/biomet/31.1-2.20
21. Keuls, M.: The use of "studentized range" in connection with an analysis of variance. Euphytica **1**(2), 112–122 (1952). https://doi.org/10.1007/bf01908269
22. Scheffe, H.: The Analysis of Variance. Wiley, New York (1999)
23. Bohrer, R.: On sharpening Scheffe bounds. J. Roy. Stat. Soc. **29**(1), 110–114 (1967)
24. Tukey, J.: Comparing individual means in analysis of variance. Biometrics **5**(2), 99–114 (1949)
25. Gurvich, V., Naumova, M.: Logical contradictions in the one-way ANOVA and Tukey-Kramer multiple comparisons test with more than two groups of observations. Symmetry **13**(8), 1387 (2021)
26. Lee, S., Lee, D.K.: What is the proper way to apply the multiple comparison test? Korean J. Nesthesiol. **71**(5), 353–360 (2018)
27. Goeman, J.J.: Comparing three groups. Am. Stat. (2021). https://doi.org/10.1080/00031305.2021.2002188
28. Benjamini, Y., Hochber, Y.: Controlling the false discovery rate: a practical and powerful approach to multiple testing. J. R. Stat. Soc. Ser. B. **57**(1), 289–300 (1995)
29. McDonald, J.H.: Handbook of Biological Statistics, 3rd edn. Sparky House Publishing, Baltimore (2014)
30. Kruskal, W., Wallis, A.: Use of ranks in one-criterion variance analysis. J. Am. Stat. Assoc. **45**(260), 583–621 (1952)
31. Corder, G.W.: Foreman, D.I.: Nonparametric Statistics for Non-Statisticians, pp. 99–105, Wiley, New York (2009)
32. Agresti, A.: Categorical Data Analysis. Wiley, New York (1990)
33. Choi, W., Lee., J.W., Huh, M.H., Kang, S.H.: An algorithm for computing the exact distribution of the Kruskal-Wallis test. Commun. Stat. Simul. Comput. **32**(4), 1029–1040 (2003). https://doi.org/10.1081/SAC-129923876

# Chapter 9
# Preparing for Natural and Manmade Disasters: Pandemics, Fire, Earthquake, Flood, Hostage Situations, etc. Can AI Ease the Job?

**Abstract** The apocalyptic images from 2020 COVID-19 pandemic are not something that we will be forgetting soon. One thing is sure, we must be prepared at all times for different types of disasters whether they are natural or manmade. Are hospital managers ready? Can they decide under pressure? This chapter answers the question: can AI ease their job when it comes to disaster?

After Prometheus had stolen fire from Olympus and conferred it to the mortals, Zeus prepared a little surprise for humankind. He asked Hephaestus to create a woman out of earth. This woman received different gifts from the other gods. Her name was Pandora. Pandora travelled around with a jar, or a box as most people know, that contained misery and evil. Pandora was sent by Zeus to marry Epimetheus, Prometheus' brother. After marrying Epimetheus, Pandora opened her box, and thus all evils flew out and conquered Earth. Hesiod's version of the story stated that Hope was left inside the box, and the lid had shut it down before she could escape. "Hope for the best and prepare for the worst" is a quote by Maya Angelou. When faced with uncertainty and difficulty, we turn to hope. Hope gives us strength, courage, and helps us not to give up. But hope is not enough. To survive the worst, one must be prepared. What does "prepared" mean? It means that we should acknowledge the fact that things might go the wrong way, it means that we should be able to anticipate the challenges and difficulties that might appear along the way and be able to mitigate their impact. Prepared means to have a backup plan, and to quickly use it to respond if things do not go our way.

Do not confuse being prepared with pessimism. It is being realistic. If we prepare for the worst possible scenario, our outcomes might be improved. Luckily for us, we can use history to try to predict the future. By predicting the future, we can either avoid the catastrophe, or at least minimize its impact. By having contingency plans, we can navigate through difficult times.

We should understand that life is unpredictable, and our predictions might be off due to certain factors that are outside of our control. In this chapter we are going to discuss a special type of AI algorithms named time series.

 
S. Belciug, *The Hospital Manager's Guide to Artificial Intelligence*, Intelligent Systems Reference Library 263, https://doi.org/10.1007/978-3-031-80314-7_9

The data represented by time series are sequences of data resulting from measurements that follow a deterministic order. So far, most statistical analyses presented in this book discussed random samples of observations. In the case of time series, the analyses are based on the assumption that successive values in the dataset represent consecutive measurements taken at equal time intervals. These temporal measurements depict a process dynamic in time. Basically, when we are discussing time series we have two goals in mind:

1. To identify the nature of the phenomenon represented by the sequence of observations that are being analyzed. This implies that we should be able to understand the underlying process that generates the data. We aim to discover patterns, trends, and behaviors, so that ultimately to be able to determine whether the phenomenon is deterministic, stochastic, or a merger between them, and whether it follows a trend or a periodicity as time passes.
2. Being able to forecast future values of the variable representing the time series using a priori known data. This means predicting upcoming points in the sequence. In this process, we consider historical trends, patterns, correlations within the data, in order to estimate the variable's future behavior, and assess the likely evolution of the studied phenomenon.

Processing and analyzing time series data is totally different from other data. Why? Because time series data has an order and it is chronological. So, we can say that the data has a temporal dimension and is naturally ordered. These facts make the data dependent and linked over time, and not identically distributed. We need to take into consideration the fact that there might be different underlying relationships, or feedback loops, or other effects. Another characteristic of time series data is that it is stochastic, and not deterministic, and obviously dynamic. These facts make it nonstationary, nonlinear, and heteroscedastic. Outliers and missing data can also be encountered. Time series data needs to be preprocessed before being analyzed, [1, 2].

Let's see what steps are needed to be followed before we start to analyze it:

1. First you need to define the problem you are dealing with, and the objective you need to reach. What is the goal of the task? Do you want to explain the phenomenon, why it behaves the way it does, predict what might happen in the future, detect anomalies, compare different phenomenon, or optimize the decision-making process?
2. Next, you must collect and clean the data: check whether the source is reliable or not, the format and the quality of the data. Recall the data engineering process.
3. Apply exploratory data analysis to look for patterns, seasonality, trends, or anomalies.
4. Model the data and evaluate the results.
5. Interpret the results and then communicate them.

In what follows, we shall present different methods that you can use to model time series data. We will denote a time series with $(\xi_t, t \in T)$, where $T$ represents the time. Before we start, we need to define some notions that we will be working with.

**Table 9.1** COVID-19 admission data

| Month | Admissions |
|---|---|
| January | 200 |
| February | 250 |
| March | 220 |
| April | 280 |
| May | 300 |
| June | 320 |

- The *trend* represents the tendency of the data. It shows whether the movement is upward or downward.
- The *seasonality* represents the patterns that might be spotted in the data, patterns that are repeated at fixed intervals within the time series. They can be spotted hourly, daily, weekly, monthly, etc. Different factors that cannot be controlled, such as the weather, the holidays, economics, politics, have an impact on it.
- The *noise* is represented by unpredictable data that cannot be explained by either the trend or the seasonality.

***The Moving Average (MA) Model***

When we use this type of model, we are forecasting future values based on the average past observations. Short-term fluctuations in data can be captured by this method.

*The Simple Moving Average (SMA)*

We shall explain the theory using the following example: let us suppose that a hospital manager wants to predict future COVID-19 patient admissions, to plan for resources (e.g. ICU beds and staff). In Table 9.1 we present the data from past 6 months during the pandemic.

The manager will use a $n$-month SMA to smooth the data and forecast the values for the next month.

$$SMA = \frac{P_{t-2} + P_{t-1} + P_t}{n},$$

where $P_t$ is the number of patients admitted in the month $t$, and $n$ is the number of months that we want to use for the moving average.

We need to compute the SMA for July as follows:

$$SMA_{July} = \frac{280 + 300 + 320}{3} = 300.$$

In Fig. 9.1 we have plotted the actual times series, and the values forecasted with SMA.

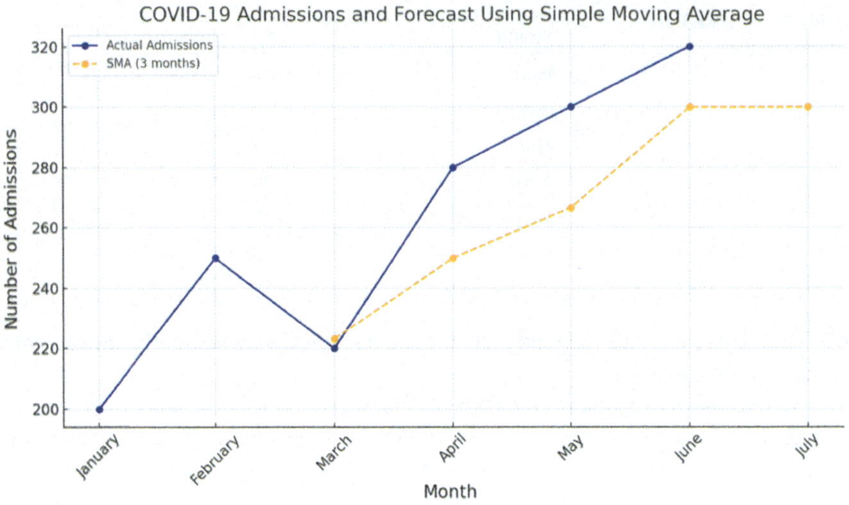


**Fig. 9.1** Forecast COVID-19 admissions using SMA

*Weighted Moving Average (WMA)*

WMA is a variation of SMA. We apply different weights to each data point, giving more weight, or importance, to recent observations. We use this method, when we believe that more recent data has greater impact or influence on the future. Let's apply WMA on our example. For 3-WMA the weights will be: 3 for the most recent month, 2 for the second most recent month, and 1 for the third most recent month.

The formula for WMA is:

$$WMA = \frac{w_1 \cdot P_t + w_2 \cdot P_{t-1} + w_3 \cdot P_{t-2}}{w_1 + w_2 + w_3}.$$

The number of admissions computed with 3-WMA for July is:

$$WMA_{July} = \frac{1 \cdot 280 + 2 \cdot 300 + 3 \cdot 320}{1 + 2 + 3} = 306.67.$$

In Fig. 9.2 we have plotted the WMA, SMA, and the admission data. In this example, we can see that WMA predicts a more responsive forecast than SMA.

*Exponential Moving Average (EMA) or Exponential Smoothing*

For this method, we will adjust a little bit the COVID-19 scenario. We will monitor the data over a 7-day period, aiming at forecasting the admissions for the 8th day. EMA is more responsive to recent trends, that can be found in rapidly changing situations, like the pandemic.

Table 9.2 presents the data for the last 7 days.

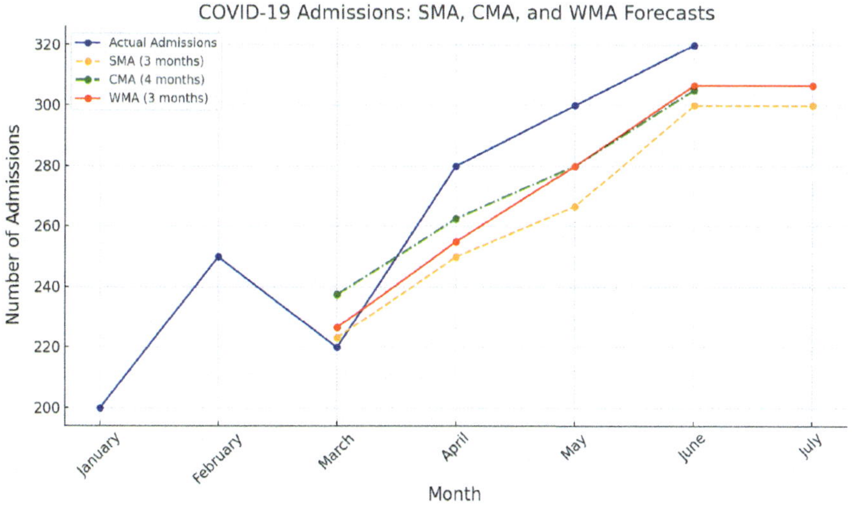


**Fig. 9.2** Comparison between WMA and SMA

**Table 9.2** COVID-19 admissions in days

| Day | Admissions |
|---|---|
| 1 | 200 |
| 2 | 250 |
| 3 | 220 |
| 4 | 280 |
| 5 | 300 |
| 6 | 320 |
| 7 | 290 |

For this example, we will use the 3-EMA to predict the admissions for day 8. The formula is:

$$EMA_t = \alpha \cdot P_t + (1 - \alpha) \cdot EMA_{t-1}$$

where $\alpha$ is the smoothing factor. Let us suppose that $\alpha = 0.5$. The initial EMA will be computed as the SMA for the first three days:

$$SMA_{day3} = \frac{200 + 250 + 220}{3} = 223.33$$

We will continue to compute EMA for each day, until we reach day 8.

$$EMA_{day4} = 0.5 \cdot 280 + (1 - 0.5) \cdot 223.33 = 251.67$$

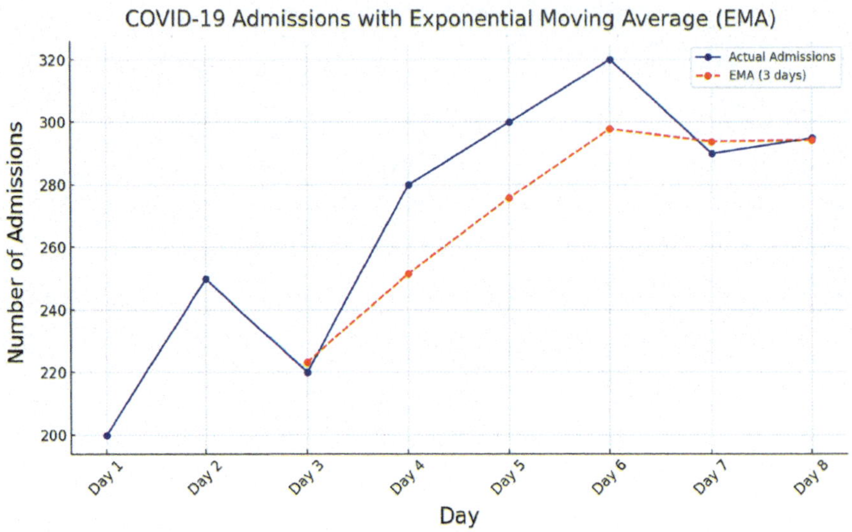


**Fig. 9.3** EMA example for COVID-19 admissions

$$EMA_{day5} = 0.5 \cdot 300 + (1 - 0.5) \cdot 251.67 = 275.84$$

$$EMA_{day6} = 0.5 \cdot 320 + (1 - 0.5) \cdot 275.84 = 297.92$$

$$EMA_{day7} = 0.5 \cdot 290 + (1 - 0.5) \cdot 297.92 = 293.96$$

$$EMA_{day8} = 0.5 \cdot 295 + (1 - 0.5) \cdot 293.96 = 294.48$$

The plot is depicted in Fig. 9.3.

If we change the value of $\alpha$ to 0.3, we will have the following forecast depicted in Fig. 9.4. This forecast is called Exponential Weighted Moving Average (EWMA).

*Double exponential smoothing method*

Let's change the situation for this method. Let us imagine that we have experienced a major earthquake, from which it resulted a high number of casualties. The hospitals are facing rapidly fluctuating patient admissions. We need to forecast the number of patients for the next hours, so that we could manage the emergency room resources and staff allocation. Table 9.3 presents the hourly patient admissions for the first 7 h after the earthquake.

We will use the Double Exponential Smoothing to predict the number of casualties for hour 8. This method is also called Holt's method. We use the following equations for level (smoothed value), trend, and forecast.

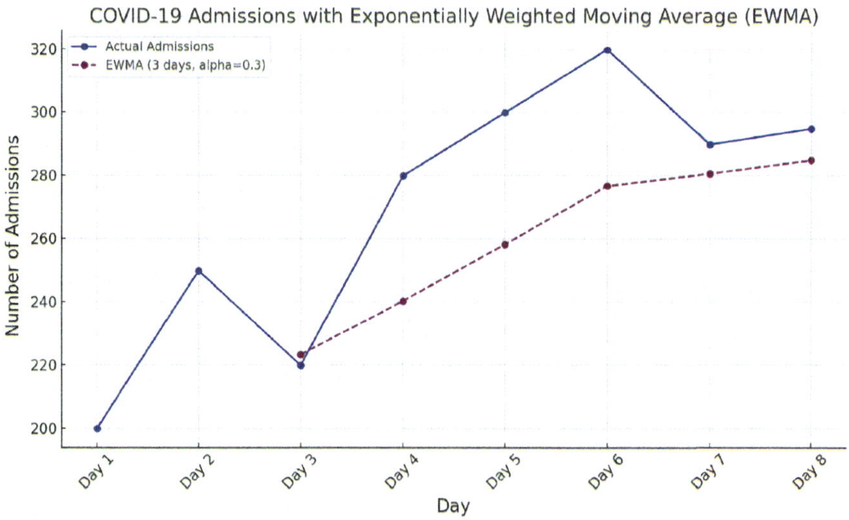


**Fig. 9.4** EWMA example for COVID-19 admissions

**Table 9.3** Hourly patient admissions

| Hour | Admissions |
|---|---|
| 1 | 50 |
| 2 | 70 |
| 3 | 60 |
| 4 | 90 |
| 5 | 80 |
| 6 | 100 |
| 7 | 95 |

- Level (smoothed value):

$$S_t = \alpha \cdot P_t + (1 - \alpha) \cdot (S_{t-1} + T_{t-1})$$

- Trend:

$$T_t = \beta \cdot (S_t - S_{t-1}) + (1 - \beta) \cdot T_{t-1}$$

- Forecast:

$$F_{t+m} = S_t + m \cdot T_t.$$

Let us use the following smoothing parameters: $\alpha = 0.5$, $\beta = 0.3$. Obviously, we could use different optimization techniques, such as the ones discussed throughout the book, to optimize these parameters in order to achieve better results.

For hour 1, we shall assume that the initial level is $S_1 = P_1 = 50$, with the initial trend $T_1 = 0$. We will start to compute the Double Exponential Smoothing for every single hour, until we reach the 8th hour.

- Hour 2:

$$S_2 = 0.5 \cdot 70 + (1 - 0.5) \cdot (50 + 0) = 35 + 25 = 60$$

$$T_2 = 0.3 \cdot (60 - 50) + 0.7 \cdot 0 = 3$$

- Hour 3:

$$S_2 = 0.5 \cdot 60 + (1 - 0.5) \cdot (60 + 3) = 30 + 31.5 = 61.5$$

$$T_2 = 0.3 \cdot (61.5 - 60) + 0.7 \cdot 0.3 = 0.45 + 2.1 = 2.55$$

- Hour 4:

$$S_4 = 0.5 \cdot 60 + (1 - 0.5) \cdot (60 + 3) = 30 + 31.5 = 61.5$$

$$T_4 = 0.3 \cdot (61.5 - 60) + 0.7 \cdot 0.3 = 0.45 + 2.1 = 2.55$$

- Hour 5:

$$S_5 = 0.5 \cdot 80 + (1 - 0.5) \cdot (77.03 + 6.45) = 40 + 41.74 = 81.74$$

$$T_5 = 0.3 \cdot (81.74 - 77.03) + 0.7 \cdot 6.45 = 1.41 + 4.52 = 5.93$$

- Hour 6:

$$S_6 = 0.5 \cdot 100 + (1 - 0.5) \cdot (81.74 + 5.93) = 50 + 43.83 = 93.83$$

$$T_6 = 0.3 \cdot (93.83 - 81.74) + 0.7 \cdot 5.93 = 3.62 + 4.15 = 7.77$$

- Hour 7:

$$S_7 = 0.5 \cdot 95 + (1 - 0.5) \cdot (93.83 + 7.77) = 47.5 + 50.8 = 98.3$$

$$T_7 = 0.3 \cdot (98.3 - 93.83) + 0.7 \cdot 7.77 = 1.34 + 5.44 = 6.78$$

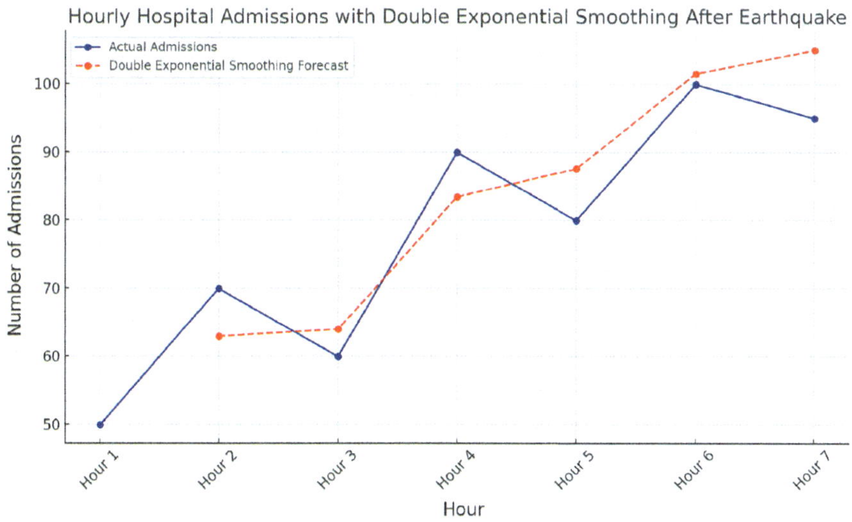


**Fig. 9.5** Prediction of casualties after major earthquake using the Double Exponential Smoothing

Now, we can forecast the number of casualties for hour 8:

$$F_8 = S_7 + T_7 = 98.3 + 6.78 = 105.08.$$

In Fig. 9.5, we present the plot of the actual admissions, and the forecast performed with the Double Exponential Smoothing.

*Triple Exponential Smoothing*

Let's model the influx of casualties after a hostage situation, in which the victims that have been released are being admitted over several hours. We need to forecast the number of casualties in the coming hours so that we manage our resources effectively. The imagined situation involves hourly trends, and seasonal patterns. Therefore, we need to apply a different method, the triple exponential smoothing or the Holt-Winters method, in which we use the level (long-term trend), the trend (rate of increase or decrease), and the seasonality (repeating patterns). Table 9.4 presents the data regarding the hourly patient admissions during the first 7 h, after the hostages started being released.

The triple exponential smoothing extends the double exponential method, by adding the seasonal component. Let's see the equations for computing all three elements:

- Level (smoothed value):

$$S_t = \alpha \cdot \frac{P_t}{I_{t-L}} + (1 - \alpha) \cdot (S_{t-1} + T_{t-1})$$

**Table 9.4** Hourly hostage situation

| Hour | Admissions |
|---|---|
| 1 | 20 |
| 2 | 35 |
| 3 | 30 |
| 4 | 50 |
| 5 | 40 |
| 6 | 60 |
| 7 | 55 |

- Trend:

$$T_t = \beta \cdot (S_t - S_{t-1}) + (1 - \beta) \cdot T_{t-1}$$

- Seasonality:

$$I_t = \gamma \cdot \frac{P_t}{S_t} + (1 - \gamma) \cdot I_{t-L}$$

- Forecast:

$$F_{t+m} = (S_t + m \cdot T_t) \cdot I_{t-L+m}$$

where $\alpha, \beta, \gamma$ are the smoothing parameters, that can be optimized using swarm intelligence, and $L$ is the length of seasonal cycle. For this example, we are going to use the following values: $\alpha = 0.4, \beta = 0.2, \gamma = 0.3, L = 3$.

The initial values for the first 3 h are the following:

$$S_1 = P_1 = 20$$

$$T_1 = 0$$

$$I_1 = \frac{P_1}{S_1}, I_2 = \frac{P_2}{S_2}, I_3 = \frac{P_3}{S_3}$$

We apply the above equations to compute the triple exponential smoothing for each hour, and then predict the influx of casualties for the 8th hour.

- Hour 4:

$$S_4 = 0.4 + \frac{50}{1} + (1 - 0.4) \cdot (30 + 0) = 38$$

because we can see that it is able to merge the seasonality, and thus can provide better results.

We are going to close this chapter with an important forecasting method: Autoregressive Integrated Moving Average or ARIMA.

*Autoregressive Integrated Moving Average (ARIMA)*

Just like you have gotten used to, we cannot apply certain methods, unless we verify some prerequisites first. In ARIMA's case, we need to check whether our time series data is stationary. Stationary means that the mean, variance and autocorrelation of our data does not change over time. To verify this, we can either use visual methods or statistical tests. If the data fails to be stationary, we can use differencing to transform it to being stationary. Differencing means subtracting the current value from the previous values of from a lagged value. In this way we can eliminate the trend and seasonality from the data, while reducing its complexity. *Lag* refers to the time difference between two observations (i.e. lag 1 refers to the relationship between an observation and the precedent observation, lag 2 refers to the relationship between an observation and the one that occurred two-time steps before, etc.), [3].

Before starting to perform ARIMA, one must specify the interval at which the data was collected (e.g. frequency—hourly, daily, etc.), and the time period one wishes to analyze or forecast.

There are multiple ARIMA order models, so after verifying the above assumptions, we need to select the optimal order model. We denote the order by three parameters $(p, d, q)$, where $p$ (AR) is the order of the autoregressive term, $d$ is the differencing, and $q$ (MA) is the order of the moving average.

The order of differencing is determined while we verify the data, whereas the $p$ and $q$ can be determined using:

- The *autocorrelation function* (ACF). By plotting the ACF we can see whether the data is correlated with itself at different lags. Using ACF we can identify the MA term. If we see on the graph a sharp cut-off after a lag, then that indicates a possible MA term of that order. Figure 9.8 shows such a plot. We can see how the autocorrelation values are declining. We see a sharp cut-off at lag 2, suggesting a MA(2) term for ARIMA. MA(2) means that the errors at lag 2 are significant, but become insignificant beyond that.
- *Partial autocorrelation function* (PACF): this plot depicts the autocorrelation of the time series data at different lags, but after we removed the effects of previous lags. PACF helps us identify the AR term. A sharp cut-off in the PACF indicates a possible AR term at that lag. Figure 9.9 shows a plot of a PACF.

We can see a cut-off at lag 1, so this could indicate a AR(1) term, and another at lag 3, which can indicate an AR(3).

- *Akaike information criterion* (AIC) measures the goodness-of-fit of the model. If a model is too complex with too many parameters, it is penalized by AIC. If we have multiple models, we can use AIC to compare and select the optimal one, with different orders of $p$ and $q$.

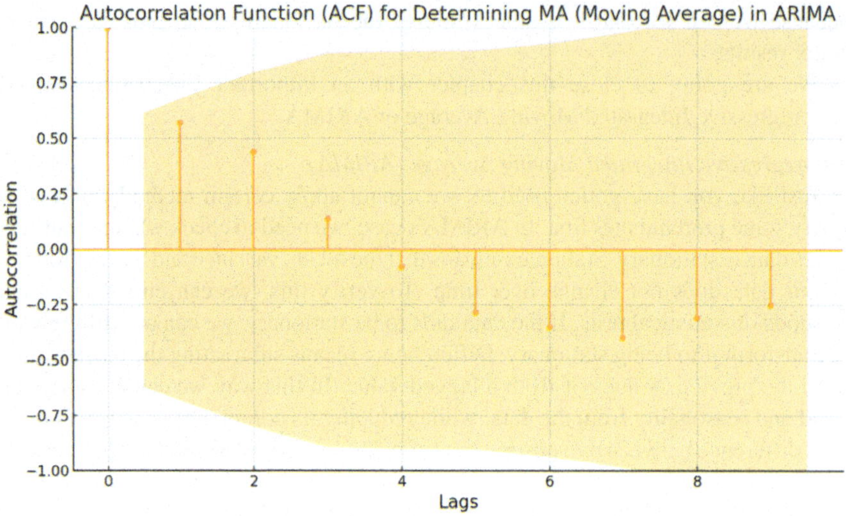


**Fig. 9.8** ACF function plot to determine AM

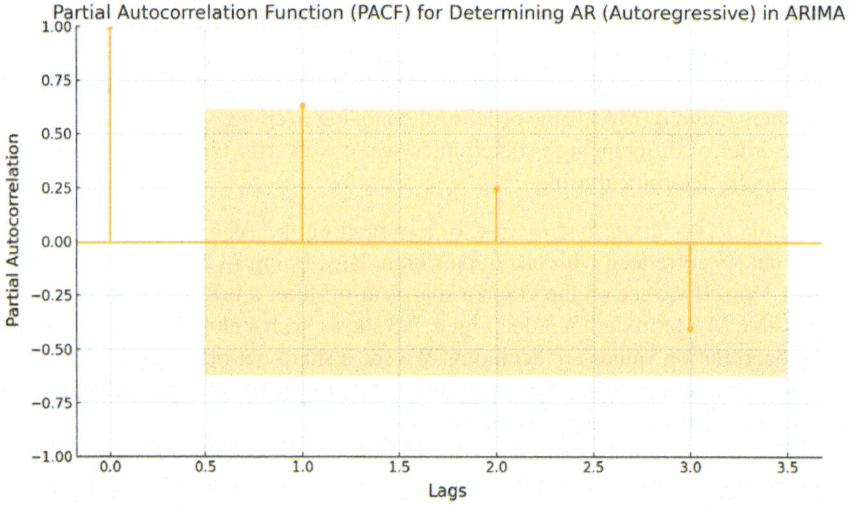


**Fig. 9.9** PACF function plot to determine AR

- *Bayesian information criterion* (BIC) also measures the goodness-of-fit of the model. BIC gives a higher penalty to models that have too many parameters.

Let us imagine the following scenario: we are running a hospital, and we are trying to forecast the number of admissions after an attack at a music concert. Multiple casualties are announced. We are going to use ARIMA to manage the resources: medical

**Table 9.5** Admissions after attack at concert

| Hours | Admissions |
|---|---|
| 1 | 100 |
| 2 | 130 |
| 3 | 150 |
| 4 | 160 |
| 5 | 180 |
| 6 | 170 |
| 7 | 200 |
| 8 | 190 |
| 9 | 210 |
| 10 | 220 |

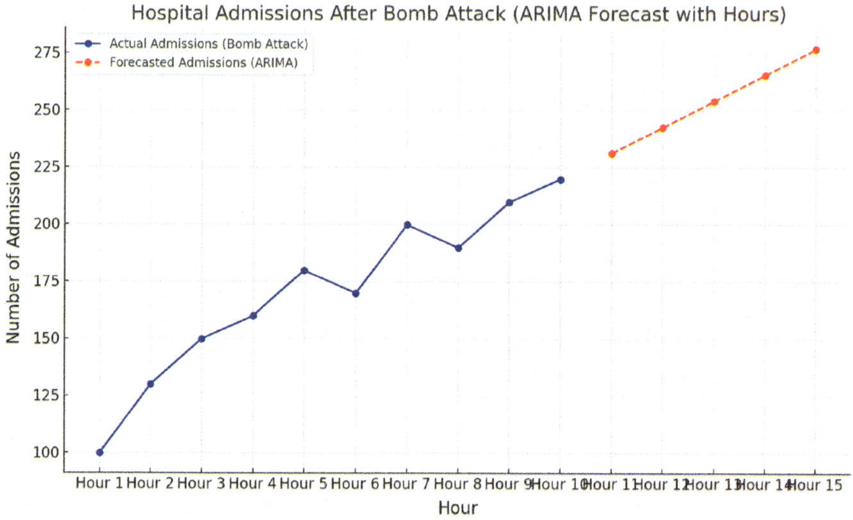


**Fig. 9.10** ARIMA(1,1,1) model

staff, supplies, and equipment. Table 9.5 presents the daily hospital admissions over the course of 10 h after the attack.

Figure 9.10 presents the ARIMA(1, 1, 1) forecast plot for admissions. The blue line represents the first 10 h, and the forecasted values for the next 5 h are depicted with red. The prediction shows that the admissions will continue to rise over the next 5 h. Even if the number is still rising, we can see that the increase is done at a slower rate compared to the initial hours.

**Table 9.6** Collected data

| Days | Admissions |
|---|---|
| 1 | 80 |
| 2 | 90 |
| 3 | 100 |
| 4 | 110 |
| 5 | 130 |
| 6 | 120 |
| 7 | 140 |
| 8 | 150 |
| 9 | 160 |
| 10 | 170 |
| 11 | 180 |
| 12 | 190 |

We have used the ARIMA(1, 1, 1) because:

- AR(1): we use the previous hour's admissions to predict the current hour value (lag 1).
- I(1): we had to difference once the data to make it stationary (e.g. removing the trends).
- MA(1): we use lag 1 error (the difference between the observed and predicted values).

Let us see another example where we are dealing with admissions after a flood (Table 9.6). We will use ARIMA (2,0,2):

- AR(2): we are using the last two hours of admissions to predict the future values
- I(0): the data is stationary, so there is no need for us to use differencing
- MA(2): we use the errors from the last 2 h for prediction adjustment.

Figure 9.11 presents the ARIMA (2,0,2) plot. We can see that the number of patients will increase, but the increase will be slow.

We have arrived at the end of this chapter. We have studied multiple models that allow us to manage the resources in case of natural or manmade disasters. Hopefully, the examples used in this chapter will help you model other situations.

On this note, we have reached the final destination of our journey together. We hope you have enjoyed watching how the hospital management revolution can take place. By using AI, we can make healthcare accessible to everyone, everywhere. As we have seen throughout this book, AI is able to provide all this. The aim of this book was to show you how we can use AI to make a difference in hospital management. There are still numerous things that need to be resolved, but we should take everything one step at a time. Till we meet again, dear reader, we hope this book will provide the right answers to your questions.

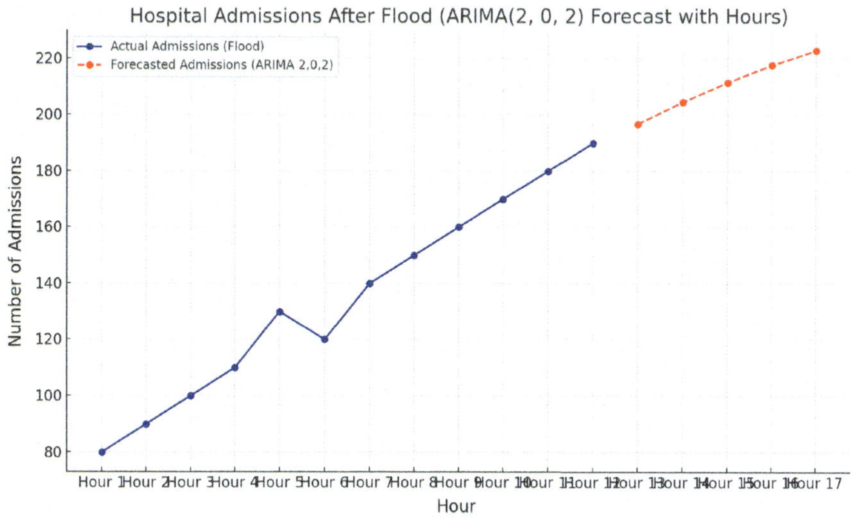


**Fig. 9.11** ARIMA (2,0,2) for admissions after flood

# References

1. Shumway, R.H, Stoffer, D.S.: Time Series Analysis and its Applications: With R Examples. Springer (2017)
2. Box, G.E.P., Gwilym, M, Reinsel, G.C., Ljung, G.M.: Time Series Analysis: Forecasting and Control, 5th edn. Wiley (2016)
3. Schaffer, A.L., Dobbins, T.A., Pearson, S.A.: Interrupted time series analysis using autoregressive integrated moving average (ARIMA models): a guide for evaluating large-scale health interventions. BMC Med. Res. Methodol. **21**(1), 58 (2021)

# Appendix—Statistical Tables

The purpose of this appendix is to present the tables used for computing the results shown in this book. It will guide you in finding the critical values and determining the appropriate *p*-levels for various statistical tests. We chose to include this appendix after discovering numerous errors and incorrect information during an online search.

In this appendix you will find the following tables:

A1 Normal distribution—areas in one tail ($z \rightarrow p$)
A2 Two-tailed $z$ test
A3 Kolmogorov-Smirnov Goodness of Fit (K-S) test table
A4 Lilliefors test
A5 Shapiro-Wilk critical values and *p*-values
A6 Tabulated coefficients for the Shapiro-Wilk test
A7 Cramér-von Mises test critical values
A8 Anderson-Darling test critical values
A9 The $\chi^2$ distribution table (for D'Agostino Pearson test)
A10 Jarque-Bera test critical values
A11 *p*-level for the $t$ distribution (one-tailed probability)
A12 *p*-value for the $t$ distribution (two-tailed probability)
A13 $F$ distribution
A14 Wilcoxon one sample (or matched pairs) test
A15 The Mann-Whitney test (Wilcoxon two sample test)
A16 Welch $t$-test



S. Belciug, *The Hospital Manager's Guide to Artificial Intelligence*, Intelligent Systems Reference Library 263, https://doi.org/10.1007/978-3-031-80314-7

**Table A1** Normal distribution—areas in one tail ($z \rightarrow p$)

| $z$ | $p_{lower}$ | $p_{upper}$ | $z$ | $p_{lower}$ | $p_{upper}$ | $z$ | $p_{lower}$ | $p_{upper}$ |
|---|---|---|---|---|---|---|---|---|
| 0.00 | 0.5000 | 0.5000 | | | | | | |
| 0.01 | 0.5040 | 0.4960 | 1.08 | 0.8599 | 0.1401 | 2.15 | 0.9842 | 0.0158 |
| 0.02 | 0.5080 | 0.4920 | 1.09 | 0.8621 | 0.1379 | 2.16 | 0.9846 | 0.0154 |
| 0.03 | 0.5120 | 0.4880 | 1.10 | 0.8643 | 0.1357 | 2.17 | 0.9850 | 0.0150 |
| 0.04 | 0.5160 | 0.4840 | 1.11 | 0.8665 | 0.1335 | 2.18 | 0.9854 | 0.0146 |
| 0.05 | 0.5199 | 0.4801 | 1.12 | 0.8686 | 0.1314 | 2.19 | 0.9857 | 0.0143 |
| 0.06 | 0.5239 | 0.4761 | 1.13 | 0.8708 | 0.1292 | 2.20 | 0.9861 | 0.0139 |
| 0.07 | 0.5279 | 0.4721 | 1.14 | 0.8729 | 0.1271 | 2.21 | 0.9864 | 0.0136 |
| 0.08 | 0.5319 | 0.4681 | 1.15 | 0.8749 | 0.1251 | 2.22 | 0.9868 | 0.0132 |
| 0.09 | 0.5359 | 0.4641 | 1.16 | 0.8770 | 0.1230 | 2.23 | 0.9871 | 0.0129 |
| 0.10 | 0.5398 | 0.4602 | 1.17 | 0.8790 | 0.1210 | 2.24 | 0.9875 | 0.0125 |
| 0.11 | 0.5438 | 0.4562 | 1.18 | 0.8810 | 0.1190 | 2.25 | 0.9878 | 0.0122 |
| 0.12 | 0.5478 | 0.4522 | 1.19 | 0.8830 | 0.1170 | 2.26 | 0.9881 | 0.0119 |
| 0.13 | 0.5517 | 0.4483 | 1.20 | 0.8849 | 0.1151 | 2.27 | 0.9884 | 0.0116 |
| 0.14 | 0.5557 | 0.4443 | 1.21 | 0.8869 | 0.1131 | 2.28 | 0.9887 | 0.0113 |
| 0.15 | 0.5596 | 0.4404 | 1.22 | 0.8888 | 0.1112 | 2.29 | 0.9890 | 0.0110 |
| 0.16 | 0.5636 | 0.4364 | 1.23 | 0.8907 | 0.1093 | 2.30 | 0.9893 | 0.0107 |
| 0.17 | 0.5675 | 0.4325 | 1.24 | 0.8925 | 0.1075 | 2.31 | 0.9896 | 0.0104 |
| 0.18 | 0.5714 | 0.4286 | 1.25 | 0.8944 | 0.1056 | 2.32 | 0.9898 | 0.0102 |
| 0.19 | 0.5753 | 0.4247 | 1.26 | 0.8962 | 0.1038 | 2.33 | 0.9901 | 0.0099 |
| 0.20 | 0.5793 | 0.4207 | 1.27 | 0.8980 | 0.1020 | 2.34 | 0.9904 | 0.0096 |
| 0.21 | 0.5832 | 0.4168 | 1.28 | 0.8987 | 0.1003 | 2.35 | 0.9906 | 0.0094 |
| 0.22 | 0.5871 | 0.4129 | 1.29 | 0.9015 | 0.0985 | 2.36 | 0.9909 | 0.0091 |
| 0.23 | 0.5910 | 0.4090 | 1.30 | 0.9032 | 0.0968 | 2.37 | 0.9911 | 0.0089 |
| 0.24 | 0.5948 | 0.4052 | 1.31 | 0.9049 | 0.0951 | 2.38 | 0.9913 | 0.0087 |
| 0.25 | 0.5987 | 0.4013 | 1.32 | 0.9066 | 0.0934 | 2.39 | 0.9916 | 0.0084 |
| 0.26 | 0.6026 | 0.3974 | 1.33 | 0.9082 | 0.0918 | 2.40 | 0.9918 | 0.0082 |
| 0.27 | 0.6064 | 0.3936 | 1.34 | 0.9099 | 0.0901 | 2.41 | 0.9920 | 0.0080 |
| 0.28 | 0.6103 | 0.3897 | 1.35 | 0.9115 | 0.0885 | 2.42 | 0.9922 | 0.0078 |
| 0.29 | 0.6141 | 0.3859 | 1.36 | 0.9131 | 0.0869 | 2.43 | 0.9925 | 0.0075 |
| 0.30 | 0.6179 | 0.3821 | 1.37 | 0.9147 | 0.0853 | 2.44 | 0.9927 | 0.0073 |
| 0.31 | 0.6217 | 0.3783 | 1.38 | 0.9162 | 0.0838 | 2.45 | 0.9929 | 0.0071 |
| 0.32 | 0.6255 | 0.3745 | 1.39 | 0.9177 | 0.0823 | 2.46 | 0.9931 | 0.0069 |
| 0.33 | 0.6293 | 0.3707 | 1.40 | 0.9192 | 0.0808 | 2.47 | 0.9932 | 0.0068 |
| 0.34 | 0.6332 | 0.3669 | 1.41 | 0.9207 | 0.0793 | 2.48 | 0.9934 | 0.0066 |
| 0.35 | 0.6368 | 0.3632 | 1.42 | 0.9222 | 0.0778 | 2.49 | 0.9936 | 0.0064 |
| 0.36 | 0.6406 | 0.3594 | 1.43 | 0.9236 | 0.0764 | 2.50 | 0.9938 | 0.0062 |

(continued)

**Table A1** (continued)

| $z$ | $p_{lower}$ | $p_{upper}$ | $z$ | $p_{lower}$ | $p_{upper}$ | $z$ | $p_{lower}$ | $p_{upper}$ |
|---|---|---|---|---|---|---|---|---|
| 0.37 | 0.6443 | 0.3557 | 1.44 | 0.9251 | 0.0749 | 2.51 | 0.9940 | 0.0060 |
| 0.38 | 0.6480 | 0.3520 | 1.45 | 0.9265 | 0.0735 | 2.52 | 0.9941 | 0.0059 |
| 0.39 | 0.6517 | 0.3483 | 1.46 | 0.9279 | 0.0721 | 2.53 | 0.9943 | 0.0057 |
| 0.40 | 0.6554 | 0.3446 | 1.47 | 0.9292 | 0.0708 | 2.54 | 0.9945 | 0.0055 |
| 0.41 | 0.6591 | 0.3409 | 1.48 | 0.9306 | 0.0694 | 2.55 | 0.9946 | 0.0054 |
| 0.42 | 0.6628 | 0.3372 | 1.49 | 0.9319 | 0.0681 | 2.56 | 0.9948 | 0.0052 |
| 0.43 | 0.6664 | 0.3336 | 1.50 | 0.9332 | 0.0668 | 2.57 | 0.9949 | 0.0051 |
| 0.44 | 0.6700 | 0.3300 | 1.51 | 0.9345 | 0.0655 | 2.58 | 0.9951 | 0.0049 |
| 0.45 | 0.6736 | 0.3264 | 1.52 | 0.9357 | 0.0643 | 2.59 | 0.9952 | 0.0048 |
| 0.46 | 0.6772 | 0.3228 | 1.53 | 0.9370 | 0.0630 | 2.60 | 0.9953 | 0.0047 |
| 0.47 | 0.6808 | 0.3192 | 1.54 | 0.9382 | 0.0618 | 2.61 | 0.9955 | 0.0045 |
| 0.48 | 0.6844 | 0.3156 | 1.55 | 0.9394 | 0.0606 | 2.62 | 0.9956 | 0.0044 |
| 0.49 | 0.6879 | 0.3121 | 1.56 | 0.9406 | 0.0594 | 2.63 | 0.9957 | 0.0043 |
| 0.50 | 0.6915 | 0.3085 | 1.57 | 0.9418 | 0.0582 | 2.64 | 0.9959 | 0.0041 |
| 0.51 | 0.6950 | 0.3050 | 1.58 | 0.9429 | 0.0571 | 2.65 | 0.9960 | 0.0040 |
| 0.52 | 0.6985 | 0.3015 | 1.59 | 0.9441 | 0.0559 | 2.66 | 0.9961 | 0.0039 |
| 0.53 | 0.7019 | 0.2981 | 1.60 | 0.9452 | 0.0548 | 2.67 | 0.9962 | 0.0038 |
| 0.54 | 0.7054 | 0.2946 | 1.61 | 0.9463 | 0.0537 | 2.68 | 0.9963 | 0.0037 |
| 0.55 | 0.7088 | 0.2912 | 1.62 | 0.9474 | 0.0526 | 2.69 | 0.9964 | 0.0036 |
| 0.56 | 0.7123 | 0.2877 | 1.63 | 0.9484 | 0.0516 | 2.70 | 0.9965 | 0.0035 |
| 0.57 | 0.7157 | 0.2843 | 1.64 | 0.9495 | 0.0505 | 2.71 | 0.9966 | 0.0034 |
| 0.58 | 0.7190 | 0.2810 | 1.65 | 0.9505 | 0.0495 | 2.72 | 0.9967 | 0.0033 |
| 0.59 | 0.7224 | 0.2776 | 1.66 | 0.9515 | 0.0485 | 2.73 | 0.9968 | 0.0032 |
| 0.60 | 0.7257 | 0.2743 | 1.67 | 0.9525 | 0.0475 | 2.74 | 0.9969 | 0.0031 |
| 0.61 | 0.7291 | 0.2709 | 1.68 | 0.9535 | 0.0465 | 2.75 | 0.9970 | 0.0030 |
| 0.62 | 0.7324 | 0.2676 | 1.69 | 0.9545 | 0.0455 | 2.76 | 0.9971 | 0.0029 |
| 0.63 | 0.7357 | 0.2643 | 1.70 | 0.9554 | 0.0446 | 2.77 | 0.9972 | 0.0028 |
| 0.64 | 0.7389 | 0.2611 | 1.71 | 0.9564 | 0.0436 | 2.78 | 0.9973 | 0.0027 |
| 0.65 | 0.7422 | 0.2578 | 1.72 | 0.9573 | 0.0427 | 2.79 | 0.9974 | 0.0026 |
| 0.66 | 0.7454 | 0.2546 | 1.73 | 0.9582 | 0.0418 | 2.80 | 0.9974 | 0.0026 |
| 0.67 | 0.7486 | 0.2514 | 1.74 | 0.9591 | 0.0409 | 2.81 | 0.9975 | 0.0025 |
| 0.68 | 0.7517 | 0.2483 | 1.75 | 0.9599 | 0.0401 | 2.82 | 0.9976 | 0.0024 |
| 0.69 | 0.7549 | 0.2451 | 1.76 | 0.9608 | 0.0392 | 2.83 | 0.9977 | 0.0023 |
| 0.70 | 0.7580 | 0.2420 | 1.77 | 0.9616 | 0.0384 | 2.84 | 0.9977 | 0.0023 |
| 0.71 | 0.7611 | 0.2389 | 1.78 | 0.9625 | 0.0375 | 2.85 | 0.9978 | 0.0022 |
| 0.72 | 0.7642 | 0.2358 | 1.79 | 0.9633 | 0.0367 | 2.86 | 0.9979 | 0.0021 |
| 0.73 | 0.7674 | 0.2327 | 1.80 | 0.9641 | 0.0359 | 2.87 | 0.9979 | 0.0021 |

(continued)

**Table A1** (continued)

| $z$ | $p_{lower}$ | $p_{upper}$ | $z$ | $p_{lower}$ | $p_{upper}$ | $z$ | $p_{lower}$ | $p_{upper}$ |
|---|---|---|---|---|---|---|---|---|
| 0.74 | 0.7704 | 0.2296 | 1.81 | 0.9649 | 0.0351 | 2.88 | 0.9980 | 0.0020 |
| 0.75 | 0.7734 | 0.2266 | 1.82 | 0.9656 | 0.0344 | 2.89 | 0.9981 | 0.0019 |
| 0.76 | 0.7764 | 0.2236 | 1.83 | 0.9664 | 0.0336 | 2.90 | 0.9981 | 0.0019 |
| 0.77 | 0.7794 | 0.2206 | 1.84 | 0.9671 | 0.0329 | 2.91 | 0.9982 | 0.0018 |
| 0.78 | 0.7823 | 0.2177 | 1.85 | 0.9678 | 0.0322 | 2.92 | 0.9982 | 0.0018 |
| 0.79 | 0.7852 | 0.2148 | 1.86 | 0.9686 | 0.0314 | 2.93 | 0.9983 | 0.0017 |
| 0.80 | 0.7881 | 0.2119 | 1.87 | 0.9693 | 0.0307 | 2.94 | 0.9984 | 0.0016 |
| 0.81 | 0.7910 | 0.2090 | 1.88 | 0.9699 | 0.0301 | 2.95 | 0.9984 | 0.0016 |
| 0.82 | 0.7939 | 0.2061 | 1.89 | 0.9706 | 0.0294 | 2.96 | 0.9985 | 0.0015 |
| 0.83 | 0.7967 | 0.2033 | 1.90 | 0.9713 | 0.0287 | 2.97 | 0.9985 | 0.0015 |
| 0.84 | 0.7995 | 0.2005 | 1.91 | 0.9719 | 0.0281 | 2.98 | 0.9986 | 0.0014 |
| 0.85 | 0.8023 | 0.1977 | 1.92 | 0.9726 | 0.0274 | 2.99 | 0.9986 | 0.0014 |
| 0.86 | 0.8051 | 0.1949 | 1.93 | 0.9732 | 0.0268 | 3.00 | 0.9987 | 0.0013 |
| 0.87 | 0.8078 | 0.1922 | 1.94 | 0.9738 | 0.0262 | 3.05 | 0.99886 | 0.00114 |
| 0.88 | 0.8106 | 0.1894 | 1.95 | 0.9744 | 0.0256 | 3.10 | 0.99903 | 0.00097 |
| 0.89 | 0.8133 | 0.1867 | 1.96 | 0.9750 | 0.0250 | 3.15 | 0.99918 | 0.00082 |
| 0.90 | 0.8159 | 0.1841 | 1.97 | 0.9756 | 0.0244 | 3.20 | 0.99931 | 0.00069 |
| 0.91 | 0.8186 | 0.1814 | 1.98 | 0.9761 | 0.0239 | 3.25 | 0.99942 | 0.00058 |
| 0.92 | 0.8212 | 0.1788 | 1.99 | 0.9767 | 0.0233 | 3.30 | 0.99952 | 0.00048 |
| 0.93 | 0.8238 | 0.1762 | 2.00 | 0.9772 | 0.0228 | 3.35 | 0.99960 | 0.00040 |
| 0.94 | 0.8264 | 0.1736 | 2.01 | 0.9778 | 0.0222 | 3.40 | 0.99966 | 0.00034 |
| 0.95 | 0.8289 | 0.1711 | 2.02 | 0.9783 | 0.0217 | 3.45 | 0.99972 | 0.00028 |
| 0.96 | 0.8315 | 0.1685 | 2.03 | 0.9788 | 0.0212 | 3.50 | 0.99977 | 0.00023 |
| 0.97 | 0.8340 | 0.1660 | 2.04 | 0.9793 | 0.0207 | 3.55 | 0.99981 | 0.00019 |
| 0.98 | 0.8365 | 0.1635 | 2.05 | 0.9798 | 0.0202 | 3.60 | 0.99984 | 0.00016 |
| 0.99 | 0.8389 | 0.1611 | 2.06 | 0.9803 | 0.0197 | 3.65 | 0.99987 | 0.00013 |
| 1.00 | 0.8413 | 0.1587 | 2.07 | 0.9808 | 0.0192 | 3.70 | 0.99989 | 0.00011 |
| 1.01 | 0.8438 | 0.1562 | 2.08 | 0.9812 | 0.0188 | 3.75 | 0.99991 | 0.00009 |
| 1.02 | 0.8461 | 0.1539 | 2.09 | 0.9817 | 0.0183 | 3.80 | 0.99993 | 0.00007 |
| 1.03 | 0.8485 | 0.1515 | 2.10 | 0.9821 | 0.0179 | 3.85 | 0.99994 | 0.00006 |
| 1.04 | 0.8508 | 0.1492 | 2.11 | 0.9826 | 0.0174 | 3.90 | 0.99995 | 0.00005 |
| 1.05 | 0.8531 | 0.1468 | 2.12 | 0.9830 | 0.0170 | 3.95 | 0.99996 | 0.00004 |
| 1.06 | 0.8554 | 0.1446 | 2.13 | 0.9834 | 0.0166 | 4.00 | 0.99997 | 0.00003 |
| 1.07 | 0.8577 | 0.1423 | 2.14 | 0.9838 | 0.0162 | | | |

**Table A5** *P*-value for the Shapiro Wilk test

| $n$ | $p$ | 0.01 | 0.02 | 0.05 | 0.1 | 0.5 | 0.9 | 0.95 | 0.98 | 0.99 |
|---|---|---|---|---|---|---|---|---|---|---|
| 3 | | 0.753 | 0.756 | 0.767 | 0.789 | 0.959 | 0.998 | 0.999 | 1.000 | 1.000 |
| 4 | | 0.687 | 0.707 | 0.748 | 0.792 | 0.935 | 0.987 | 0.992 | 0.996 | 0.997 |
| 5 | | 0.686 | 0.715 | 0.762 | 0.806 | 0.927 | 0.979 | 0.986 | 0.991 | 0.993 |
| 6 | | 0.713 | 0.743 | 0.788 | 0.826 | 0.927 | 0.974 | 0.981 | 0.986 | 0.989 |
| 7 | | 0.730 | 0.760 | 0.803 | 0.838 | 0.928 | 0.972 | 0.979 | 0.985 | 0.988 |
| 8 | | 0.749 | 0.778 | 0.818 | 0.851 | 0.932 | 0.972 | 0.978 | 0.984 | 0.987 |
| 9 | | 0.764 | 0.791 | 0.829 | 0.859 | 0.935 | 0.972 | 0.978 | 0.983 | 0.986 |
| 10 | | 0.781 | 0.806 | 0.842 | 0.869 | 0.938 | 0.972 | 0.978 | 0.983 | 0.986 |
| 11 | | 0.792 | 0.817 | 0.850 | 0.876 | 0.940 | 0.973 | 0.979 | 0.984 | 0.986 |
| 12 | | 0.805 | 0.828 | 0.859 | 0.883 | 0.943 | 0.973 | 0.979 | 0.984 | 0.986 |
| 13 | | 0.814 | 0.837 | 0.866 | 0.889 | 0.945 | 0.974 | 0.979 | 0.984 | 0.986 |
| 14 | | 0.825 | 0.846 | 0.874 | 0.895 | 0.947 | 0.975 | 0.980 | 0.984 | 0.986 |
| 15 | | 0.835 | 0.855 | 0.881 | 0.901 | 0.950 | 0.975 | 0.980 | 0.984 | 0.987 |
| 16 | | 0.844 | 0.863 | 0.887 | 0.906 | 0.952 | 0.976 | 0.981 | 0.985 | 0.987 |
| 17 | | 0.851 | 0.869 | 0.892 | 0.910 | 0.954 | 0.977 | 0.981 | 0.985 | 0.987 |
| 18 | | 0.858 | 0.874 | 0.897 | 0.914 | 0.956 | 0.978 | 0.982 | 0.986 | 0.988 |
| 19 | | 0.863 | 0.879 | 0.901 | 0.917 | 0.957 | 0.978 | 0.982 | 0.986 | 0.988 |
| 20 | | 0.868 | 0.884 | 0.905 | 0.920 | 0.959 | 0.979 | 0.983 | 0.987 | 0.989 |
| 21 | | 0.873 | 0.888 | 0.908 | 0.923 | 0.960 | 0.980 | 0.983 | 0.987 | 0.989 |
| 22 | | 0.878 | 0.892 | 0.911 | 0.926 | 0.960 | 0.980 | 0.983 | 0.987 | 0.989 |
| 23 | | 0.881 | 0.895 | 0.914 | 0.928 | 0.962 | 0.981 | 0.984 | 0.987 | 0.989 |
| 24 | | 0.884 | 0.898 | 0.916 | 0.930 | 0.963 | 0.981 | 0.984 | 0.988 | 0.989 |
| 25 | | 0.888 | 0.901 | 0.918 | 0.931 | 0.964 | 0.981 | 0.985 | 0.988 | 0.989 |
| 26 | | 0.891 | 0.904 | 0.920 | 0.933 | 0.965 | 0.982 | 0.985 | 0.988 | 0.989 |
| 27 | | 0.894 | 0.906 | 0.923 | 0.935 | 0.965 | 0.982 | 0.985 | 0.988 | 0.990 |
| 28 | | 0.896 | 0.908 | 0.924 | 0.936 | 0.966 | 0.982 | 0.985 | 0.988 | 0.990 |
| 29 | | 0.898 | 0.910 | 0.926 | 0.937 | 0.966 | 0.982 | 0.985 | 0.988 | 0.990 |
| 30 | | 0.900 | 0.912 | 0.927 | 0.939 | 0.967 | 0.983 | 0.985 | 0.988 | 0.990 |
| 31 | | 0.902 | 0.914 | 0.929 | 0.940 | 0.967 | 0.983 | 0.986 | 0.988 | 0.990 |
| 32 | | 0.904 | 0.915 | 0.030 | 0.941 | 0.968 | 0.983 | 0.986 | 0.988 | 0.990 |
| 33 | | 0.906 | 0.917 | 0.931 | 0.942 | 0.968 | 0.983 | 0.986 | 0.989 | 0.990 |
| 34 | | 0.908 | 0.919 | 0.933 | 0.943 | 0.969 | 0.983 | 0.986 | 0.989 | 0.990 |
| 35 | | 0.910 | 0.920 | 0.934 | 0.944 | 0.969 | 0.984 | 0.986 | 0.989 | 0.990 |
| 36 | | 0.912 | 0.922 | 0.935 | 0.945 | 0.970 | 0.984 | 0.986 | 0.989 | 0.990 |
| 37 | | 0.914 | 0.924 | 0.936 | 0.946 | 0.970 | 0.984 | 0.987 | 0.989 | 0.990 |
| 38 | | 0.916 | 0.925 | 0.938 | 0.947 | 0.971 | 0.984 | 0.987 | 0.989 | 0.990 |
| 39 | | 0.917 | 0.927 | 0.939 | 0.948 | 0.971 | 0.984 | 0.987 | 0.989 | 0.991 |

(continued)

**Table A5** (continued)

| *n* \| *p* | 0.01 | 0.02 | 0.05 | 0.1 | 0.5 | 0.9 | 0.95 | 0.98 | 0.99 |
|---|---|---|---|---|---|---|---|---|---|
| 40 | 0.919 | 0.928 | 0.940 | 0.949 | 0.972 | 0.985 | 0.987 | 0.989 | 0.991 |
| 41 | 0.920 | 0.929 | 0.941 | 0.950 | 0.972 | 0.985 | 0.987 | 0.989 | 0.991 |
| 42 | 0.922 | 0.930 | 0.942 | 0.950 | 0.972 | 0.985 | 0.987 | 0.989 | 0.991 |
| 43 | 0.923 | 0.932 | 0.943 | 0.951 | 0.973 | 0.985 | 0.987 | 0.990 | 0.991 |
| 44 | 0.924 | 0.933 | 0.944 | 0.952 | 0.973 | 0.985 | 0.988 | 0.990 | 0.991 |
| 45 | 0.926 | 0.934 | 0.945 | 0.953 | 0.974 | 0.985 | 0.988 | 0.990 | 0.991 |
| 46 | 0.927 | 0.935 | 0.945 | 0.953 | 0.974 | 0.985 | 0.988 | 0.990 | 0.991 |
| 47 | 0.928 | 0.936 | 0.946 | 0.954 | 0.974 | 0.985 | 0.988 | 0.990 | 0.991 |
| 48 | 0.929 | 0.937 | 0.947 | 0.954 | 0.974 | 0.985 | 0.988 | 0.990 | 0.991 |
| 49 | 0.929 | 0.938 | 0.947 | 0.955 | 0.974 | 0.985 | 0.988 | 0.990 | 0.991 |
| 50 | 0.930 | 9.939 | 0.947 | 0.955 | 0.974 | 0.985 | 0.988 | 0.990 | 0.991 |

**Table A6** Shapiro Wilk tabulated coefficients $a_i$

| $n$ | 2 | 3 | 4 | 5 | 6 | 7 | 8 | 9 | 10 | 11 | 12 | 13 | 14 |
|---|---|---|---|---|---|---|---|---|---|---|---|---|---|
| $a_1$ | 0.7071 | 0.7071 | 0.6872 | 0.6646 | 0.6431 | 0.6233 | 0.6052 | 0.5888 | 0.5739 | 0.5601 | 0.5475 | 0.5359 | 0.5251 |
| $a_2$ | | | 0.1677 | 0.2413 | 0.2805 | 0.3031 | 0.3164 | 0.3244 | 0.3291 | 0.3315 | 0.3325 | 0.3325 | 0.3318 |
| $a_3$ | | | | | 0.0875 | 0.1401 | 0.1743 | 0.1976 | 0.2141 | 0.2260 | 0.2347 | 0.2412 | 0.2460 |
| $a_4$ | | | | | | | 0.0561 | 0.0947 | 0.1224 | 0.1429 | 0.1585 | 0.1707 | 0.1802 |
| $a_5$ | | | | | | | | | 0.0399 | 0.0695 | 0.0922 | 0.1099 | 0.1240 |
| $a_6$ | | | | | | | | | | | 0.0803 | 0.0539 | 0.0727 |
| $a_7$ | | | | | | | | | | | | | 0.0240 |

| $n$ | 15 | 16 | 17 | 18 | 19 | 20 | 21 | 22 | 23 | 24 | 25 | 26 |
|---|---|---|---|---|---|---|---|---|---|---|---|---|
| $a_1$ | 0.5150 | 0.5056 | 0.4968 | 0.4886 | 0.4808 | 0.4734 | 0.4643 | 0.4590 | 0.4542 | 0.4493 | 0.4450 | 0.4407 |
| $a_2$ | 0.3306 | 0.3290 | 0.3273 | 0.3253 | 0.3232 | 0.3211 | 0.3185 | 0.3156 | 0.3126 | 0.3098 | 0.3069 | 0.3043 |
| $a_3$ | 0.2495 | 0.2521 | 0.2540 | 0.2553 | 0.2561 | 0.2565 | 0.2578 | 0.2571 | 0.2563 | 0.2554 | 0.2543 | 0.2533 |
| $a_4$ | 0.1878 | 0.1939 | 0.1988 | 0.2027 | 0.2059 | 0.2085 | 0.2119 | 0.2131 | 0.2139 | 0.2145 | 0.2148 | 0.2151 |
| $a_5$ | 0.1353 | 0.1447 | 0.1524 | 0.1587 | 0.1641 | 0.1686 | 0.1736 | 0.1764 | 0.1787 | 0.1807 | 0.1822 | 0.1836 |
| $a_6$ | 0.0880 | 0.1005 | 0.1109 | 0.1197 | 0.1271 | 0.1334 | 0.1399 | 0.1443 | 0.1480 | 0.1512 | 0.1539 | 0.1563 |
| $a_7$ | 0.0433 | 0.0593 | 0.0725 | 0.0837 | 0.0932 | 0.1013 | 0.1092 | 0.1150 | 0.1201 | 0.1245 | 0.1283 | 0.1316 |
| $a_8$ | | 0.0196 | 0.0359 | 0.0496 | 0.0612 | 0.0711 | 0.0804 | 0.0878 | 0.0941 | 0.0997 | 0.1046 | 0.1089 |
| $a_9$ | | | | 0.0163 | 0.0303 | 0.0422 | 0.0530 | 0.0618 | 0.0696 | 0.0764 | 0.0823 | 0.876 |
| $a_{10}$ | | | | | | 0.0140 | 0.0263 | 0.0368 | 0.0459 | 0.0593 | 0.0610 | 0.0672 |
| $a_{11}$ | | | | | | | | 0.0122 | 0.0228 | 0.0321 | 0.0403 | 0.0476 |
| $a_{12}$ | | | | | | | | | 0.0000 | 0.0107 | 0.0200 | 0.0284 |
| $a_{13}$ | | | | | | | | | | | 0.0000 | 0.0094 |

(continued)

**Table A6** (continued)

| $n$ | 27 | 28 | 29 | 30 | 31 | 32 | 33 | 34 | 35 | 36 | 37 | 38 |
|---|---|---|---|---|---|---|---|---|---|---|---|---|
| $a_1$ | 0.4366 | 0.4328 | 0.4291 | 0.4254 | 0.4220 | 0.4188 | 0.4156 | 0.4127 | 0.4096 | 0.4068 | 0.4040 | 0.4015 |
| $a_2$ | 0.3018 | 0.2992 | 0.2968 | 0.2944 | 0.2921 | 0.2898 | 0.2876 | 0.2854 | 0.2834 | 0.2813 | 0.2794 | 0.2774 |
| $a_3$ | 0.2522 | 0.2510 | 0.2499 | 0.2487 | 0.2475 | 0.2463 | 0.2451 | 0.2439 | 0.2427 | 0.2415 | 0.2403 | 0.2391 |
| $a_4$ | 0.2152 | 0.2151 | 0.2150 | 0.2148 | 0.2145 | 0.2141 | 0.2137 | 0.2132 | 0.2127 | 0.2121 | 0.2116 | 0.2110 |
| $a_5$ | 0.1848 | 0.1857 | 0.1864 | 0.1870 | 0.1874 | 0.1878 | 0.1880 | 0.1882 | 0.1883 | 0.1883 | 0.1883 | 0.1881 |
| $a_6$ | 0.1584 | 0.1601 | 0.1616 | 0.1630 | 0.1641 | 0.1651 | 0.1660 | 0.1667 | 0.1673 | 0.1678 | 0.1683 | 0.1686 |
| $a_7$ | 0.1346 | 0.1372 | 0.1395 | 0.1415 | 0.1433 | 0.1449 | 0.1463 | 0.1475 | 0.1487 | 0.1496 | 0.1505 | 0.1513 |
| $a_8$ | 0.1128 | 0.1162 | 0.1192 | 0.1219 | 0.1243 | 0.1265 | 0.1284 | 0.1301 | 0.1317 | 0.1331 | 0.1344 | 0.1356 |
| $a_9$ | 0.0923 | 0.0965 | 0.1002 | 0.1036 | 0.1066 | 0.1093 | 0.1118 | 0.1140 | 0.1160 | 0.1179 | 0.1196 | 0.1211 |
| $a_{10}$ | 0.0728 | 0.0778 | 0.0822 | 0.0862 | 0.0899 | 0.0931 | 0.0961 | 0.0988 | 0.1013 | 0.1036 | 0.1056 | 0.1075 |
| $a_{11}$ | 0.0540 | 0.0598 | 0.0650 | 0.697 | 0.0739 | 0.0777 | 0.0812 | 0.0844 | 0.0873 | 0.0900 | 0.0924 | 0.0947 |
| $a_{12}$ | 0.0358 | 0.0424 | 0.0483 | 0.0537 | 0.0585 | 0.0629 | 0.0669 | 0.0706 | 0.0739 | 0.0770 | 0.0798 | 0.0824 |
| $a_{13}$ | 0.0178 | 0.0253 | 0.0320 | 0.0381 | 0.0435 | 0.0485 | 0.0530 | 0.0572 | 0.0610 | 0.0645 | 0.0677 | 0.0706 |
| $a_{14}$ | 0.0000 | 0.0084 | 0.0159 | 0.0227 | 0.0289 | 0.0344 | 0.0395 | 0.0441 | 0.0484 | 0.0523 | 0.0559 | 0.0592 |
| $a_{15}$ | | | 0.0000 | 0.0076 | 0.0144 | 0.0206 | 0.0262 | 0.0314 | 0.0361 | 0.0404 | 0.0444 | 0.0481 |
| $a_{16}$ | | | | | 0.0000 | 0.0068 | 0.0131 | 0.0187 | 0.0239 | 0.0287 | 0.0331 | 0.0372 |
| $a_{17}$ | | | | | | | 0.0000 | 0.0062 | 0.0119 | 0.0172 | 0.0220 | 0.0264 |
| $a_{18}$ | | | | | | | | | 0.0000 | 0.0057 | 0.0110 | 0.0158 |
| $a_{19}$ | | | | | | | | | | | 0.0000 | 0.0053 |

(continued)

**Table A6** (continued)

| $n$ | 39 | 40 | 41 | 42 | 43 | 44 | 45 | 46 | 47 | 48 | 49 | 50 |
|---|---|---|---|---|---|---|---|---|---|---|---|---|
| $a_1$ | 0.3989 | 0.3964 | 0.3940 | 0.3917 | 0.3894 | 0.3872 | 0.3850 | 0.3830 | 0.3808 | 0.3789 | 0.3770 | 0.3751 |
| $a_2$ | 0.2755 | 0.2737 | 0.2719 | 0.2701 | 0.2684 | 0.2667 | 0.2651 | 0.2635 | 0.2620 | 0.2604 | 0.2589 | 0.2574 |
| $a_3$ | 0.2380 | 0.2368 | 0.2357 | 0.2345 | 0.2334 | 0.2323 | 0.2313 | 0.2302 | 0.2291 | 0.2281 | 0.2271 | 0.2260 |
| $a_4$ | 0.2104 | 0.2098 | 0.2091 | 0.2085 | 0.2078 | 0.2072 | 0.2065 | 0.2058 | 0.2052 | 0.2045 | 0.2038 | 0.2032 |
| $a_5$ | 0.1880 | 0.1878 | 0.1876 | 0.1874 | 0.1871 | 0.1868 | 0.1865 | 0.1862 | 0.1859 | 0.1855 | 0.1851 | 0.1847 |
| $a_6$ | 0.1689 | 0.1691 | 0.1693 | 0.1694 | 0.1695 | 0.1695 | 0.1695 | 0.1695 | 0.1695 | 0.1693 | 0.1692 | 0.1691 |
| $a_7$ | 0.1520 | 0.1526 | 0.1531 | 0.1535 | 0.1539 | 0.1542 | 0.1545 | 0.1548 | 0.1550 | 0.1551 | 0.1553 | 0.1554 |
| $a_8$ | 0.1366 | 0.1376 | 0.1384 | 0.1392 | 0.1398 | 0.1405 | 0.1410 | 0.1415 | 0.1420 | 0.1423 | 0.1427 | 0.1430 |
| $a_9$ | 0.1225 | 0.1237 | 0.1249 | 0.1259 | 0.1269 | 0.1278 | 0.1286 | 0.1293 | 0.1300 | 0.1306 | 0.1312 | 0.1317 |
| $a_{10}$ | 0.1092 | 0.1108 | 0.1123 | 0.1136 | 0.1149 | 0.1160 | 0.1170 | 0.1180 | 0.1189 | 0.1197 | 0.1205 | 0.1212 |
| $a_{11}$ | 0.0967 | 0.0986 | 0.1003 | 0.1020 | 0.1035 | 0.1049 | 0.1062 | 0.1073 | 0.1085 | 0.1095 | 0.1105 | 0.1113 |
| $a_{12}$ | 0.0848 | 0.0870 | 0.0891 | 0.0909 | 0.0927 | 0.0943 | 0.0959 | 0.0972 | 0.0986 | 0.0998 | 0.1010 | 0.1020 |
| $a_{13}$ | 0.0733 | 0.0759 | 0.0782 | 0.0804 | 0.0824 | 0.0842 | 0.0860 | 0.0876 | 0.0892 | 0.0906 | 0.0919 | 0.0932 |
| $a_{14}$ | 0.0622 | 0.0651 | 0.0677 | 0.0701 | 0.0724 | 0.0745 | 0.0765 | 0.0783 | 0.0801 | 0.0817 | 0.0832 | 0.0846 |
| $a_{15}$ | 0.0515 | 0.0546 | 0.0575 | 0.0602 | 0.0628 | 0.0651 | 0.0673 | 0.0694 | 0.0713 | 0.0731 | 0.0748 | 0.0764 |
| $a_{16}$ | 0.0409 | 0.0444 | 0.0476 | 0.0506 | 0.0534 | 0.0560 | 0.0584 | 0.0607 | 0.0628 | 0.0648 | 0.0667 | 0.0685 |
| $a_{17}$ | 0.0305 | 0.0343 | 0.0379 | 0.0411 | 0.0442 | 0.0471 | 0.0497 | 0.0522 | 0.0546 | 0.0568 | 0.0588 | 0.0608 |
| $a_{18}$ | 0.0203 | 0.0244 | 0.0283 | 0.0318 | 0.0352 | 0.0383 | 0.0412 | 0.0439 | 0.0465 | 0.0489 | 0.0511 | 0.0532 |
| $a_{19}$ | 0.0101 | 0.0146 | 0.0188 | 0.0227 | 0.0263 | 0.0296 | 0.0328 | 0.0357 | 0.0385 | 0.0411 | 0.0436 | 0.0459 |
| $a_{20}$ | 0.0000 | 0.0049 | 0.0094 | 0.0136 | 0.0175 | 0.0211 | 0.0245 | 0.0277 | 0.0307 | 0.0335 | 0.0361 | 0.0386 |
| $a_{21}$ | | | 0.0000 | 0.0045 | 0.0087 | 0.0126 | 0.0163 | 0.0197 | 0.0229 | 0.0259 | 0.288 | 0.0314 |
| $a_{22}$ | | | | | 0.0000 | 0.0042 | 0.0081 | 0.0118 | 0.0153 | 0.0185 | 0.0215 | 0.244 |

(continued)

**Table A6** (continued)

| $n$ | 39 | 40 | 41 | 42 | 43 | 44 | 45 | 46 | 47 | 48 | 49 | 50 |
|---|---|---|---|---|---|---|---|---|---|---|---|---|
| $a_{23}$ | | | | | | | 0.0000 | 0.0039 | 0.0076 | 0.0111 | 0.0143 | 0.0174 |
| $a_{24}$ | | | | | | | | | 0.0000 | 0.0037 | 0.0071 | 0.0104 |
| $a_{25}$ | | | | | | | | | | | 0.000 | 0.0037 |

**Table A7** Critical values for Cramér-von Mises test

| $n$ | $p$-level | | | | |
|---|---|---|---|---|---|
| | 0.20 | 0.15 | 0.10 | 0.05 | 0.01 |
| 2 | 0.138 | 0.149 | 0.162 | 0.175 | 0.186 |
| 3 | 0.121 | 0.135 | 0.154 | 0.184 | 0.23 |
| 4 | 0.121 | 0.134 | 0.155 | 0.191 | 0.28 |
| 5 | 0.123 | 0.137 | 0.160 | 0.199 | 0.30 |
| 6 | 0.124 | 0.139 | 0.162 | 0.204 | 0.31 |
| 7 | 0.124 | 0.140 | 0.165 | 0.208 | 0.32 |
| 8 | 0.124 | 0.141 | 0.165 | 0.210 | 0.32 |
| 9 | 0.125 | 0.142 | 0.167 | 0.212 | 0.32 |
| 10 | 0.125 | 0.142 | 0.167 | 0.212 | 0.32 |
| 11 | 0.126 | 0.144 | 0.169 | 0.214 | 0.32 |
| 12 | 0.126 | 0.144 | 0.169 | 0.214 | 0.32 |
| 13 | 0.126 | 0.144 | 0.169 | 0.214 | 0.33 |
| 14 | 0.126 | 0.144 | 0.169 | 0.214 | 0.33 |
| 15 | 0.126 | 0.144 | 0.169 | 0.215 | 0.33 |
| 16 | 0.127 | 0.145 | 0.171 | 0.216 | 0.33 |
| 17 | 0.127 | 0.145 | 0.171 | 0.217 | 0.33 |
| 18 | 0.127 | 0.146 | 0.171 | 0.217 | 0.33 |
| 19 | 0.127 | 0.146 | 0.171 | 0.217 | 0.33 |
| 20 | 0.128 | 0.146 | 0.172 | 0.217 | 0.33 |
| 30 | 0.128 | 0.146 | 0.172 | 0.218 | 0.33 |
| 60 | 0.128 | 0.147 | 0.173 | 0.220 | 0.33 |
| 100 | 0.129 | 0.147 | 0.173 | 0.220 | 0.34 |

**Table A8** Critical values for the Anderson–Darling test

| No | $n$ | 0.15 | 0.10 | 0.05 | 0.025 | 0.01 |
|---|---|---|---|---|---|---|
| 0 | $\geq 5$ | 1.621 | 1.933 | 2.492 | 3.070 | 3.878 |
| 1 | | | 0.908 | 1.105 | 1.304 | 1.573 |
| 2 | $\geq 5$ | | 1.760 | 2.323 | 2.904 | 3.690 |
| 3 | 10 | 0.514 | 0.578 | 0.693 | 0.779 | 0.926 |
| 4 | 20 | 0.528 | 0.591 | 0.704 | 0.815 | 0.969 |
| 5 | 50 | 0.546 | 0.616 | 0.735 | 0.861 | 1.021 |
| 6 | 100 | 0.559 | 0.631 | 0.754 | 0.884 | 1.047 |
| 7 | $\infty$ | 0.576 | 0.656 | 0.787 | 0.918 | 1.092 |

**Table A9** The $\chi^2$ distribution table (for D'Agostino Pearson test)

| Degrees of freedom | 0.2 | 0.1 | 0.05 | 0.02 | 0.01 | 0.001 |
|---|---|---|---|---|---|---|
| 1 | 1.642 | 2.706 | 3.841 | 5.412 | 6.635 | 10.827 |
| 2 | 3.219 | 4.605 | 5.991 | 7.824 | 9.210 | 13.815 |
| 3 | 4.642 | 6.251 | 7.815 | 9.837 | 11.345 | 16.268 |
| 4 | 5.989 | 7.770 | 9.488 | 11.688 | 13.277 | 18.465 |
| 5 | 7.289 | 9.236 | 11.070 | 13.388 | 15.086 | 20.517 |
| 6 | 8.558 | 10.645 | 12.592 | 15.033 | 16.812 | 22.457 |
| 7 | 9.803 | 12.017 | 14.067 | 16.622 | 18.475 | 24.322 |
| 8 | 11.030 | 13.362 | 15.507 | 18.168 | 20.090 | 26.125 |
| 9 | 12.242 | 14.684 | 16.919 | 19.670 | 21.666 | 27.877 |
| 10 | 13.442 | 15.987 | 18.307 | 21.161 | 23.209 | 29.588 |
| 11 | 14.631 | 17.275 | 19.675 | 22.618 | 24.725 | 31.264 |
| 12 | 15.812 | 18.549 | 21.026 | 24.054 | 26.217 | 32.909 |
| 13 | 16.985 | 19.812 | 22.362 | 25.472 | 27.688 | 34.528 |
| 14 | 18.151 | 21.064 | 23.685 | 26.873 | 29.141 | 36.123 |
| 15 | 19.311 | 22.307 | 23.685 | 26.873 | 29.141 | 36.123 |
| 16 | 20.465 | 23.542 | 26.296 | 26.633 | 32.000 | 39.252 |
| 17 | 21.615 | 24.769 | 27.587 | 30.995 | 33.408 | 40.790 |
| 18 | 22.760 | 25.989 | 28.869 | 32.346 | 34.805 | 42.312 |
| 19 | 23.900 | 27.204 | 30.144 | 22.687 | 36.191 | 43.820 |
| 20 | 25.038 | 28.412 | 31.410 | 35.020 | 37.566 | 45.315 |
| 21 | 26.171 | 29.615 | 32.671 | 36.343 | 38.932 | 46.797 |
| 22 | 27.301 | 30.813 | 33.924 | 37.659 | 40.289 | 48.268 |
| 23 | 28.429 | 32.007 | 35.172 | 38.968 | 41.638 | 49.728 |
| 24 | 29.553 | 33.196 | 36.415 | 40.270 | 42.980 | 51.179 |
| 25 | 30.675 | 34.382 | 37.652 | 41.566 | 44.314 | 52.620 |

**Table A10** Critical values of Jarque–Bera test

| $\alpha$ | 0.01 | 0.02 | 0.05 | 0.10 | 0.20 |
|---|---|---|---|---|---|
| $n = 10$ | 5.738 | 4.274 | 2.535 | 1.618 | 1.126 |
| $n = 20$ | 9.458 | 6.583 | 3.768 | 2.335 | 1.556 |
| $n = 50$ | 12.331 | 8.721 | 5.004 | 3.192 | 2.122 |
| $n = 100$ | 12.296 | 9.089 | 5.448 | 3.643 | 2.474 |
| $n = 200$ | 11.750 | 8.788 | 5.728 | 4.081 | 2.748 |
| $n = 500$ | 10.601 | 8.349 | 5.825 | 4.324 | 2.985 |
| $n \to \infty$ | 9.210 | 7.824 | 5.991 | 4.605 | 3.219 |

**Table A11** *P*-level for the *t* distribution (one-tailed probability)

| *df* | *p*-level (one-tailed probability) | | | | | | | |
|---|---|---|---|---|---|---|---|---|
| | 0.40 | 0.25 | 0.10 | 0.05 | 0.025 | 0.01 | 0.005 | 0.0005 |
| 1 | 0.324 | 1.000 | 3.077 | 6.313 | 12.706 | 31.820 | 63.656 | 636.619 |
| 2 | 0.288 | 0.816 | 1.885 | 2.919 | 4.302 | 6.964 | 9.924 | 31.599 |
| 3 | 0.276 | 0.764 | 1.637 | 2.353 | 3.182 | 4.540 | 5.840 | 12.924 |
| 4 | 0.270 | 0.740 | 1.533 | 2.131 | 2.776 | 3.746 | 4.604 | 8.610 |
| 5 | 0.267 | 0.726 | 1.475 | 2.015 | 2.570 | 3.364 | 4.032 | 6.868 |
| 6 | 0.264 | 0.717 | 1.439 | 1.943 | 2.446 | 3.142 | 3.707 | 5.958 |
| 7 | 0.263 | 0.711 | 1.414 | 1.894 | 2.364 | 2.997 | 3.499 | 5.407 |
| 8 | 0.261 | 0.706 | 1.396 | 1.859 | 2.306 | 2.896 | 3.355 | 5.014 |
| 9 | 0.260 | 0.702 | 1.383 | 1.833 | 2.262 | 2.841 | 3.249 | 4.780 |
| 10 | 0.2595 | 0.669 | 1.372 | 1.812 | 2.288 | 2.763 | 3.169 | 4.586 |
| 11 | 0.2590 | 0.697 | 1.363 | 1.795 | 2.200 | 2.718 | 3.105 | 4.437 |
| 12 | 0.2585 | 0.695 | 1.356 | 1.782 | 2.178 | 2.681 | 3.054 | 4.317 |
| 13 | 0.2582 | 0.693 | 1.350 | 1.770 | 2.160 | 2.650 | 3.012 | 4.220 |
| 14 | 0.2578 | 0.692 | 1.345 | 1.761 | 2.144 | 2.624 | 2.976 | 4.140 |
| 15 | 0.2575 | 0.691 | 1.340 | 1.753 | 2.131 | 2.602 | 2.946 | 4.072 |
| 16 | 0.2573 | 0.690 | 1.336 | 1.745 | 2.119 | 2.583 | 2.920 | 4.015 |
| 17 | 0.2571 | 0.689 | 1.333 | 1.739 | 2.109 | 2.566 | 2.898 | 3.965 |
| 18 | 0.2569 | 0.688 | 1.330 | 1.734 | 2.100 | 2.552 | 2.878 | 3.921 |
| 19 | 0.2567 | 0.687 | 1.327 | 1.729 | 2.093 | 2.539 | 2.860 | 3.883 |
| 20 | 0.2565 | 0.6869 | 1.325 | 1.724 | 2.085 | 2.527 | 2.845 | 3.883 |
| 21 | 0.2564 | 0.6863 | 1.323 | 1.720 | 2.079 | 2.517 | 2.831 | 3.849 |
| 22 | 0.2562 | 0.6858 | 1.321 | 1.717 | 2.073 | 2.508 | 2.818 | 3.819 |
| 23 | 0.2561 | 0.6853 | 1.319 | 1.713 | 2.068 | 2.499 | 2.807 | 3.792 |
| 24 | 0.2560 | 0.6848 | 1.317 | 1.710 | 2.063 | 2.492 | 2.796 | 3.745 |
| 25 | 0.2559 | 0.6844 | 1.316 | 1.708 | 2.059 | 2.485 | 2.787 | 3.725 |
| 26 | 0.2558 | 0.6840 | 1.314 | 1.705 | 2.055 | 2.478 | 2.778 | 3.706 |
| 27 | 0.2557 | 0.6836 | 1.313 | 1.703 | 2.051 | 2.472 | 2.770 | 3.689 |
| 28 | 0.25568 | 0.6833 | 1.312 | 1.701 | 2.048 | 2.467 | 2.763 | 3.673 |
| 29 | 0.25560 | 0.6827 | 1.311 | 1.699 | 2.045 | 2.462 | 2.756 | 3.659 |
| 30 | 0.255 | 0.6744 | 1.310 | 1.697 | 2.042 | 2.457 | 2.750 | 3.646 |
| *z* | 0.253 | 0.674 | 0.128 | 1.644 | 1.959 | 2.326 | 2.575 | 3.290 |

**Table A12** *P*-value for the *t* distribution (two-tailed probability)

| Degrees of freedom | Two-tailed probability (*p*) | | | | | |
|---|---|---|---|---|---|---|
| | 0.2 | 0.1 | 0.05 | 0.02 | 0.01 | 0.001 |
| 1 | 3.078 | 6.314 | 12.706 | 31.821 | 63.657 | 636.619 |
| 2 | 1.886 | 2.920 | 4.303 | 6.965 | 9.925 | 31.599 |
| 3 | 1.638 | 2.353 | 3.182 | 4.541 | 5.841 | 12.924 |
| 4 | 1.533 | 2.132 | 2.776 | 3.747 | 4.604 | 8.610 |
| 5 | 1.476 | 2.015 | 2.571 | 3.365 | 4.032 | 6.869 |
| 6 | 1.440 | 1.943 | 2.447 | 3.143 | 3.707 | 5.959 |
| 7 | 1.415 | 1.895 | 2.365 | 2.998 | 3.499 | 5.408 |
| 8 | 1.397 | 1.860 | 2.306 | 2.896 | 3.355 | 5.041 |
| 9 | 1.383 | 1.833 | 2.262 | 2.821 | 3.250 | 4.781 |
| 10 | 1.372 | 1.812 | 2.228 | 2.764 | 3.169 | 4.587 |
| 11 | 1.363 | 1.796 | 2.201 | 2.718 | 3.106 | 4.437 |
| 12 | 1.356 | 1.782 | 2.179 | 2.681 | 3.055 | 4.318 |
| 13 | 1.350 | 1.771 | 2.160 | 2.650 | 3.012 | 4.221 |
| 14 | 1.345 | 1.761 | 2.145 | 2.624 | 2.977 | 4.140 |
| 15 | 1.341 | 1.753 | 2.131 | 2.602 | 2.947 | 4.073 |
| 16 | 1.337 | 1.746 | 2.120 | 2.583 | 2.921 | 4.015 |
| 17 | 1.333 | 1.740 | 2.110 | 2.567 | 2.898 | 3.965 |
| 18 | 1.330 | 1.734 | 2.101 | 2.552 | 2.878 | 3.922 |
| 19 | 1.328 | 1.729 | 2.093 | 2.539 | 2.861 | 3.883 |
| 20 | 1.325 | 1.725 | 2.086 | 2.528 | 2.845 | 3.850 |
| 21 | 1.323 | 1.721 | 2.080 | 2.518 | 2.831 | 3.819 |
| 22 | 1.323 | 1.717 | 2.074 | 2.508 | 2.819 | 3.792 |
| 23 | 1.319 | 1.714 | 2.069 | 2.500 | 2.807 | 3.768 |
| 24 | 1.318 | 1.711 | 2.064 | 2.492 | 2.797 | 3.745 |
| 25 | 1.316 | 1.708 | 2.060 | 2.485 | 2.787 | 3.725 |
| 26 | 1.315 | 1.706 | 2.056 | 2.479 | 2.779 | 3.707 |
| 27 | 1.314 | 1.703 | 2.052 | 2.473 | 2.771 | 3.690 |
| 28 | 1.313 | 1.701 | 2.048 | 2.467 | 2.763 | 3.674 |
| 29 | 1.311 | 1.699 | 2.045 | 2.462 | 2.756 | 3.659 |
| 30 | 1.310 | 1.697 | 2.042 | 2.457 | 2.750 | 3.646 |
| 31 | 1.309 | 1.696 | 2.040 | 2.453 | 2.744 | 3.633 |
| 32 | 1.309 | 1.694 | 2.037 | 2.449 | 2.738 | 3.622 |
| 33 | 1.308 | 1.692 | 2.035 | 2.445 | 2.733 | 3.611 |
| 34 | 1.307 | 1.691 | 2.032 | 2.441 | 2.728 | 3.601 |
| 35 | 1.306 | 1.690 | 2.030 | 2.438 | 2.724 | 3.591 |
| 36 | 1.306 | 1.688 | 2.028 | 2.434 | 2.719 | 3.582 |

(continued)

**Table A12** (continued)

| Degrees of freedom | Two-tailed probability ($p$) | | | | | |
|---|---|---|---|---|---|---|
| | 0.2 | 0.1 | 0.05 | 0.02 | 0.01 | 0.001 |
| 37 | 1.305 | 1.687 | 2.026 | 2.431 | 2.715 | 3.574 |
| 38 | 1.304 | 1.686 | 2.024 | 2.429 | 2.712 | 3.566 |
| 39 | 1.304 | 1.685 | 2.023 | 2.426 | 2.708 | 3.558 |
| 40 | 1.303 | 1.684 | 2.021 | 2.423 | 2.704 | 3.551 |
| 41 | 1.303 | 1.683 | 2.020 | 2.421 | 2.701 | 3.544 |
| 42 | 1.302 | 1.682 | 2.018 | 2.418 | 2.698 | 3.538 |
| 43 | 1.302 | 1.681 | 2.017 | 2.416 | 2.695 | 3.532 |
| 44 | 1.301 | 1.680 | 2.015 | 2.414 | 2.692 | 3.526 |
| 45 | 1.301 | 1.679 | 2.014 | 2.412 | 2.690 | 3.520 |
| 46 | 1.300 | 1.679 | 2.013 | 2.410 | 2.687 | 3.515 |
| 47 | 1.300 | 1.678 | 2.012 | 2.408 | 2.685 | 3.510 |
| 48 | 1.299 | 1.677 | 2.011 | 2.407 | 2.682 | 3.505 |
| 49 | 1.299 | 1.677 | 2.010 | 2.403 | 2.680 | 3.500 |
| 50 | 1.299 | 1.676 | 2.009 | 2.403 | 2.678 | 3.496 |
| 51 | 1.298 | 1.675 | 2.008 | 2.402 | 2.676 | 3.492 |
| 52 | 1.298 | 1.675 | 2.007 | 2.400 | 2.674 | 3.488 |
| 53 | 1.297 | 1.674 | 2.006 | 2.399 | 2.672 | 3.484 |
| 54 | 1.297 | 1.674 | 2.005 | 2.397 | 2.670 | 3.480 |
| 55 | 1.297 | 1.673 | 2.004 | 2.396 | 2.668 | 2.476 |
| 56 | 1.297 | 1.673 | 2.003 | 2.395 | 2.667 | 3.473 |
| 57 | 1.297 | 1.672 | 2.002 | 2.394 | 2.665 | 3.470 |
| 58 | 1.296 | 1.672 | 2.002 | 2.392 | 2.663 | 3.466 |
| 59 | 1.296 | 1.671 | 2.001 | 2.391 | 2.662 | 3.463 |
| 60 | 1.296 | 1.671 | 2.000 | 2.390 | 2.660 | 2.460 |
| 70 | 1.294 | 1.667 | 1.994 | 2.381 | 2.648 | 3.435 |
| 80 | 1.292 | 1.667 | 1.994 | 2.381 | 2.648 | 3.435 |
| 90 | 1.292 | 1.664 | 1.990 | 2.374 | 2.639 | 3.416 |
| 100 | 1.290 | 1.660 | 1.984 | 2.364 | 2.626 | 3.390 |
| 110 | 1.289 | 1.659 | 1.982 | 2.361 | 2.621 | 3.381 |
| 120 | 1.289 | 1.658 | 1.980 | 2.358 | 2.617 | 3.373 |
| 130 | 1.288 | 1.657 | 1.978 | 2.355 | 2.614 | 3.367 |
| 140 | 1.288 | 1.656 | 1.977 | 2.353 | 2.611 | 3.361 |
| 150 | 1.287 | 1.655 | 1.976 | 2.351 | 2.609 | 3.357 |

**Table A13** *F* distribution

| $n_2$ | $p$ \ $n_1$ | 1 | 2 | 3 | 4 | 5 | 6 | 7 | 8 | 9 | 10 | 12 | 15 | 20 | ∞ |
|---|---|---|---|---|---|---|---|---|---|---|---|---|---|---|---|
| 1 | 0.1 | 39.9 | 49.5 | 53.6 | 55.8 | 57.2 | 58.2 | 59.1 | 59.7 | 60.1 | 60.5 | 61.0 | 61.5 | 62.0 | 63.3 |
| | 0.05 | 161.4 | 199.5 | 215.8 | 224.7 | 230.4 | 234.2 | 237.0 | 239.1 | 240.8 | 242.1 | 244.2 | 246.2 | 248.3 | 254.3 |
| | 0.01 | 4051.8 | 4999.5 | 5403.5 | 5624.8 | 5763.8 | 5859.2 | 5928.6 | 5981.3 | 6022.7 | 6056.1 | 6106.6 | 6157.6 | 6209.0 | 6265.9 |
| 2 | 0.1 | 8.53 | 9.00 | 9.16 | 9.24 | 9.29 | 9.33 | 9.35 | 9.37 | 9.38 | 9.39 | 9.41 | 9.43 | 9.44 | 9.49 |
| | 0.05 | 18.51 | 19.00 | 19.16 | 19.25 | 19.30 | 19.33 | 19.35 | 19.37 | 19.38 | 19.40 | 19.41 | 19.43 | 19.45 | 19.50 |
| | 0.01 | 98.50 | 99.00 | 99.17 | 99.25 | 99.30 | 99.33 | 99.36 | 99.75 | 99.78 | 99.80 | 99.83 | 99.87 | 99.90 | 99.50 |
| 3 | 0.1 | 5.54 | 5.46 | 5.39 | 5.34 | 5.31 | 5.28 | 5.27 | 5.25 | 5.24 | 5.23 | 5.22 | 5.20 | 5.18 | 5.13 |
| | 0.05 | 10.13 | 9.55 | 9.28 | 9.12 | 9.01 | 8.94 | 8.89 | 8.85 | 8.81 | 8.79 | 8.74 | 8.70 | 8.66 | 8.53 |
| | 0.01 | 34.11 | 30.82 | 29.46 | 28.71 | 28.24 | 27.91 | 27.67 | 27.49 | 27.34 | 27.23 | 27.05 | 26.87 | 26.69 | 26.13 |
| 4 | 0.1 | 4.54 | 4.32 | 4.19 | 4.11 | 4.05 | 4.01 | 3.98 | 3.95 | 3.94 | 3.92 | 3.90 | 3.87 | 3.84 | 3.76 |
| | 0.05 | 7.71 | 6.94 | 6.59 | 6.39 | 6.26 | 6.16 | 6.09 | 6.04 | 6.00 | 5.96 | 5.91 | 5.86 | 5.80 | 5.63 |
| | 0.01 | 21.20 | 18.00 | 16.69 | 15.98 | 15.52 | 15.21 | 14.98 | 14.80 | 14.66 | 14.55 | 14.37 | 14.20 | 14.02 | 13.46 |
| 5 | 0.01 | 4.06 | 3.78 | 3.62 | 3.52 | 3.45 | 3.40 | 3.37 | 3.34 | 3.32 | 3.30 | 3.27 | 3.24 | 3.21 | 3.10 |
| | 0.05 | 6.61 | 5.79 | 5.41 | 5.19 | 5.05 | 4.95 | 4.88 | 4.82 | 4.77 | 4.74 | 4.68 | 4.62 | 4.56 | 4.36 |
| | 0.01 | 16.26 | 13.27 | 12.06 | 11.39 | 10.97 | 10.67 | 10.46 | 10.29 | 10.16 | 10.05 | 9.89 | 9.72 | 9.55 | 9.02 |
| 6 | 0.1 | 3.78 | 3.46 | 3.29 | 3.18 | 3.11 | 3.05 | 3.01 | 2.98 | 2.96 | 2.94 | 2.90 | 2.87 | 2.84 | 2.72 |
| | 0.05 | 5.99 | 5.14 | 4.76 | 4.53 | 4.39 | 4.28 | 4.21 | 4.15 | 4.10 | 4.06 | 4.00 | 3.94 | 3.94 | 3.67 |
| | 0.01 | 13.74 | 10.92 | 9.78 | 9.15 | 8.75 | 8.47 | 8.26 | 8.10 | 7.98 | 7.87 | 7.72 | 7.56 | 7.56 | 6.88 |
| 7 | 0.1 | 3.59 | 3.26 | 3.07 | 2.96 | 2.88 | 2.83 | 2.78 | 2.75 | 2.72 | 2.70 | 2.67 | 2.63 | 2.59 | 2.47 |
| | 0.05 | 5.59 | 4.74 | 4.35 | 4.12 | 3.97 | 3.87 | 3.79 | 3.73 | 3.68 | 3.64 | 3.57 | 3.51 | 3.44 | 3.23 |
| | 0.01 | 12.25 | 9.55 | 8.45 | 7.85 | 7.46 | 7.19 | 6.99 | 6.84 | 6.72 | 6.62 | 6.47 | 6.31 | 6.16 | 5.65 |

(continued)

**Table A13** (continued)

| $n_1$ | | | | | | | | | | | | | | | |
|---|---|---|---|---|---|---|---|---|---|---|---|---|---|---|---|
| $n_2$ | $p$ | 1 | 2 | 3 | 4 | 5 | 6 | 7 | 8 | 9 | 10 | 12 | 15 | 20 | ∞ |
| 8 | 0.1 | 3.46 | 3.11 | 2.92 | 2.81 | 2.73 | 2.67 | 2.62 | 2.59 | 2.56 | 2.54 | 2.50 | 2.46 | 2.43 | 2.29 |
| | 0.05 | 5.32 | 4.46 | 4.07 | 3.84 | 3.69 | 3.58 | 3.50 | 3.44 | 3.39 | 3.35 | 3.28 | 3.22 | 3.15 | 2.93 |
| | 0.01 | 11.26 | 8.65 | 7.59 | 7.01 | 6.63 | 6.37 | 6.18 | 6.03 | 5.91 | 5.81 | 5.67 | 5.52 | 5.36 | 4.86 |
| 9 | 0.1 | 3.36 | 3.01 | 2.81 | 2.69 | 2.61 | 2.55 | 2.51 | 2.47 | 2.44 | 2.42 | 2.38 | 2.34 | 2.30 | 2.16 |
| | 0.05 | 5.12 | 4.26 | 3.86 | 3.63 | 3.48 | 3.37 | 3.29 | 3.23 | 3.18 | 3.14 | 3.07 | 3.01 | 2.94 | 2.71 |
| | 0.01 | 10.56 | 8.02 | 6.99 | 6.42 | 6.06 | 5.80 | 5.61 | 5.47 | 5.35 | 5.26 | 5.11 | 4.96 | 4.81 | 4.31 |
| 10 | 0.1 | 3.29 | 2.92 | 2.73 | 2.61 | 2.52 | 2.46 | 2.41 | 2.38 | 2.35 | 2.32 | 2.28 | 2.24 | 2.20 | 2.01 |
| | 0.05 | 4.96 | 4.10 | 3.71 | 3.48 | 3.33 | 3.22 | 3.13 | 3.07 | 3.02 | 2.98 | 2.84 | 2.84 | 2.77 | 2.54 |
| | 0.01 | 10.04 | 7.56 | 6.55 | 5.99 | 5.64 | 5.39 | 5.20 | 5.06 | 4.94 | 4.85 | 4.71 | 4.56 | 4.81 | 3.91 |
| 12 | 0.1 | 3.18 | 2.81 | 2.61 | 2.48 | 2.39 | 2.33 | 2.28 | 2.24 | 2.21 | 2.19 | 2.15 | 2.10 | 2.06 | 1.90 |
| | 0.05 | 4.75 | 3.89 | 3.49 | 3.26 | 3.11 | 3.00 | 2.91 | 2.85 | 2.80 | 2.75 | 2.69 | 2.62 | 2.54 | 2.30 |
| | 0.01 | 9.33 | 6.93 | 5.95 | 5.41 | 5.06 | 4.82 | 4.64 | 4.50 | 4.39 | 4.30 | 4.16 | 4.01 | 3.86 | 3.36 |
| 15 | 0.1 | 3.07 | 2.70 | 2.49 | 2.36 | 2.27 | 2.21 | 2.16 | 2.12 | 2.09 | 2.06 | 2.02 | 1.97 | 1.92 | 1.76 |
| | 0.05 | 4.54 | 3.68 | 3.29 | 3.06 | 2.90 | 2.79 | 2.71 | 2.64 | 2.59 | 2.54 | 2.48 | 2.40 | 2.33 | 2.07 |
| | 0.01 | 8.68 | 6.36 | 5.42 | 4.89 | 4.56 | 4.32 | 4.14 | 4.00 | 3.89 | 3.80 | 3.67 | 3.52 | 3.37 | 2.87 |
| 20 | 0.1 | 2.97 | 2.59 | 2.38 | 2.25 | 2.16 | 2.09 | 2.04 | 2.00 | 1.96 | 1.94 | 1.89 | 1.84 | 1.79 | 1.61 |
| | 0.05 | 4.35 | 3.49 | 2.92 | 2.87 | 2.71 | 2.60 | 2.51 | 2.45 | 2.39 | 2.35 | 2.28 | 2.20 | 2.12 | 1.84 |
| | 0.01 | 8.10 | 5.85 | 4.94 | 4.43 | 4.10 | 3.87 | 3.56 | 3.56 | 3.46 | 2.37 | 3.23 | 3.09 | 2.94 | 2.42 |
| 30 | 0.1 | 2.88 | 2.49 | 2.28 | 2.14 | 2.05 | 1.98 | 1.93 | 1.88 | 1.85 | 1.82 | 1.77 | 1.72 | 1.67 | 1.46 |
| | 0.05 | 4.17 | 3.32 | 2.92 | 2.69 | 2.53 | 2.42 | 2.33 | 2.27 | 2.21 | 2.16 | 2.09 | 2.01 | 1.93 | 1.62 |

(continued)

**Table A13** (continued)

| $n_2$ | $p$ ($n_1$) | 1 | 2 | 3 | 4 | 5 | 6 | 7 | 8 | 9 | 10 | 12 | 15 | 20 | ∞ |
|---|---|---|---|---|---|---|---|---|---|---|---|---|---|---|---|
| | 0.01 | 7.56 | 5.39 | 4.51 | 4.02 | 3.70 | 3.47 | 3.30 | 3.17 | 3.07 | 2.98 | 2.84 | 2.70 | 2.55 | 2.01 |
| 40 | 0.1 | 2.84 | 2.44 | 2.23 | 2.09 | 2.00 | 1.93 | 1.87 | 1.83 | 1.79 | 1.76 | 1.71 | 1.66 | 1.61 | 1.38 |
| | 0.05 | 4.08 | 3.23 | 2.84 | 2.61 | 2.45 | 2.34 | 2.23 | 2.18 | 2.12 | 2.08 | 2.00 | 1.92 | 1.84 | 1.51 |
| | 0.01 | 7.31 | 5.18 | 4.31 | 3.83 | 3.51 | 3.29 | 3.12 | 2.99 | 2.89 | 2.80 | 2.66 | 2.52 | 2.37 | 1.80 |
| 60 | 0.1 | 2.79 | 2.39 | 2.18 | 2.04 | 1.95 | 1.87 | 1.82 | 1.77 | 1.74 | 1.71 | 1.66 | 1.60 | 1.54 | 1.29 |
| | 0.05 | 4.00 | 3.15 | 2.76 | 2.53 | 2.37 | 2.25 | 2.17 | 2.10 | 2.04 | 1.99 | 1.92 | 1.84 | 1.75 | 1.39 |
| | 0.01 | 7.08 | 4.98 | 4.13 | 3.65 | 3.34 | 3.12 | 2.95 | 2.82 | 2.72 | 2.63 | 2.50 | 2.35 | 2.20 | 1.60 |
| 120 | 0.1 | 2.75 | 2.35 | 2.13 | 1.99 | 1.90 | 1.82 | 1.77 | 1.72 | 1.68 | 1.65 | 1.60 | 1.54 | 1.48 | 1.19 |
| | 0.05 | 3.92 | 3.07 | 2.68 | 2.45 | 2.29 | 2.18 | 2.09 | 2.02 | 1.96 | 1.91 | 1.83 | 1.75 | 1.66 | 1.25 |
| | 0.01 | 6.85 | 4.79 | 3.95 | 3.48 | 3.17 | 2.96 | 2.79 | 2.66 | 2.56 | 2.47 | 2.34 | 2.19 | 2.03 | 1.38 |
| ∞ | 0.1 | 2.71 | 2.30 | 2.08 | 1.94 | 1.85 | 1.77 | 1.72 | 1.67 | 1.63 | 1.60 | 1.55 | 1.49 | 1.42 | 1.13 |
| | 0.05 | 3.84 | 3.00 | 2.60 | 2.37 | 2.21 | 2.10 | 2.01 | 1.94 | 1.88 | 1.83 | 1.75 | 1.67 | 1.57 | 1.17 |
| | 0.01 | 6.63 | 4.61 | 3.78 | 3.32 | 3.02 | 2.80 | 2.64 | 2.51 | 2.41 | 2.32 | 2.18 | 2.04 | 1.88 | 1.24 |

**Table A14** Wilcoxon one sample (or matched pairs) test

| *n* | *p*-value | | | | | |
|---|---|---|---|---|---|---|
| | 0.2 | 0.1 | 0.05 | 0.02 | 0.01 | 0.001 |
| 4 | 0–10 | – | – | – | – | – |
| 5 | 2–13 | 0–15 | – | – | – | – |
| 6 | 3–18 | 2–19 | 0–21 | – | – | – |
| 7 | 5–23 | 3–25 | 2–26 | 0–28 | – | – |
| 8 | 8–28 | 5–31 | 3–33 | 1–35 | 0–36 | – |
| 9 | 10–35 | 8–37 | 5–40 | 3–42 | 1–44 | – |
| 10 | 14–41 | 10–45 | 8–47 | 5–50 | 3–52 | – |
| 11 | 17–49 | 13–53 | 10–56 | 7–59 | 5–61 | 0–66 |
| 12 | 21–57 | 17–61 | 13–65 | 9–69 | 7–71 | 1–77 |
| 13 | 26–65 | 21–70 | 17–74 | 12–79 | 9–82 | 2–89 |
| 14 | 31–74 | 25–80 | 21–84 | 15–90 | 12–93 | 4–101 |
| 15 | 36–84 | 30–90 | 25–95 | 19–101 | 15–105 | 6–114 |
| 16 | 42–94 | 35–101 | 29–107 | 23–113 | 19–117 | 9–127 |
| 17 | 48–105 | 41–112 | 34–119 | 28–125 | 23–130 | 11–142 |
| 18 | 55–116 | 47–12 | 40–131 | 32–139 | 27–144 | 14–157 |
| 19 | 62–128 | 53–137 | 46–144 | 37–153 | 32–158 | 18–172 |
| 20 | 69–141 | 60–150 | 52–158 | 43–167 | 37–173 | 21–189 |
| 21 | 77–154 | 67–164 | 58–173 | 49–182 | 42–189 | 26–205 |
| 22 | 86–167 | 75–178 | 66–187 | 55–198 | 48–105 | 30–223 |
| 23 | 95–181 | 83–193 | 73–203 | 62–214 | 54–222 | 35–241 |
| 24 | 104–196 | 91–209 | 81–219 | 69–231 | 61–239 | 40–260 |
| 25 | 114–211 | 100–225 | 89–239 | 76–249 | 68–257 | 45–280 |

**Table A15** The Mann–Whitney test (Wilcoxon two sample test)

| | | Two-tailed probability (P) | | | | |
|---|---|---|---|---|---|---|
| $n_1$ | $n_2$ | 0.1 | 0.05 | 0.02 | 0.01 | 0.001 |
| 3 | 3 | 6–15 | – | – | – | – |
| 3 | 4 | 6–18 | – | – | – | – |
| 4 | 4 | 11–25 | 10–26 | – | – | – |
| 2 | 5 | 3–13 | – | – | – | – |
| 3 | 5 | 7–20 | 6–21 | – | – | – |
| 4 | 5 | 12–28 | 11–29 | 10–30 | – | – |
| 5 | 5 | 19–36 | 17–38 | 16–39 | 15–40 | – |
| 2 | 6 | 3–15 | – | – | – | – |
| 3 | 6 | 8–22 | 7–23 | – | – | – |
| 4 | 6 | 13–31 | 12–32 | 11–33 | 10–34 | – |
| 5 | 6 | 20–40 | 18–42 | 17–43 | 16–44 | – |
| 6 | 6 | 28–50 | 26–52 | 24–54 | 23–55 | – |
| 2 | 7 | 3–17 | – | – | – | – |
| 3 | 7 | 8–25 | 7–26 | 6–27 | – | – |
| 4 | 7 | 14–34 | 13–35 | 11–37 | 10–38 | – |
| 5 | 7 | 21–44 | 20–45 | 18–47 | 16–49 | – |
| 6 | 7 | 29–55 | 27–57 | 25–59 | 24–60 | – |
| 7 | 7 | 39–66 | 36–69 | 34–71 | 32–73 | 28–77 |
| 2 | 8 | 4–18 | 3–19 | – | – | – |
| 3 | 8 | 9–27 | 8–28 | 6–30 | – | – |
| 4 | 8 | 15–37 | 14–38 | 12–40 | 11–41 | – |
| 5 | 8 | 23–47 | 21–49 | 19–51 | 17–53 | – |
| 6 | 8 | 31–59 | 29–61 | 27–63 | 25–65 | 21–69 |
| 7 | 8 | 41–71 | 38–74 | 35–77 | 34–78 | 29–83 |
| 8 | 8 | 51–85 | 49–87 | 45–91 | 43–93 | 38–98 |
| 2 | 9 | 4–20 | 3–21 | – | – | – |
| 3 | 9 | 10–29 | 8–31 | 7–32 | 6–33 | – |
| 4 | 9 | 16–40 | 14–42 | 13–43 | 11–45 | – |
| 5 | 9 | 24–51 | 22–53 | 20–55 | 18–57 | 15–60 |
| 6 | 9 | 33–63 | 31–65 | 28–68 | 26–70 | 22–74 |
| 7 | 9 | 43–76 | 40–79 | 37–82 | 35–84 | 30–89 |
| 8 | 9 | 54–90 | 51–93 | 47–97 | 45–99 | 40–104 |
| 9 | 9 | 66–105 | 62–109 | 59–112 | 56–115 | 50–121 |
| 2 | 10 | 4–22 | 3–23 | – | – | – |
| 3 | 10 | 10–32 | 9–33 | 7–35 | 6–36 | – |
| 4 | 10 | 17–43 | 15–45 | 13–47 | 12–48 | – |

(continued)

**Table A15** (continued)

| | | Two-tailed probability (P) | | | | |
|---|---|---|---|---|---|---|
| $n_1$ | $n_2$ | 0.1 | 0.05 | 0.02 | 0.01 | 0.001 |
| 5 | 10 | 26–54 | 23–57 | 21–59 | 19–61 | 15–65 |
| 6 | 10 | 35–67 | 32–70 | 29–73 | 27–75 | 23–79 |
| 7 | 10 | 45–81 | 42–84 | 39–87 | 37–89 | 31–95 |
| 8 | 10 | 56–96 | 53–99 | 49–103 | 47–105 | 41–111 |
| 9 | 10 | 69–111 | 65–115 | 61–119 | 58–122 | 52–128 |
| 10 | 10 | 82–128 | 78–132 | 74–136 | 71–139 | 63–147 |
| 2 | 11 | 4–24 | 3–25 | – | – | – |
| 3 | 11 | 11–34 | 9–36 | 7–38 | 6–39 | – |
| 4 | 11 | 18–46 | 16–48 | 14–50 | 12–52 | – |
| 5 | 11 | 27–58 | 24–61 | 22–63 | 20–65 | 16–69 |
| 6 | 11 | 37–71 | 34–74 | 30–78 | 28–80 | 23–85 |
| 7 | 11 | 47–86 | 44–89 | 40–93 | 38–95 | 32–101 |
| 8 | 11 | 59–101 | 55–105 | 51–109 | 49–111 | 42–118 |
| 9 | 11 | 72–117 | 68–121 | 63–126 | 61–128 | 53–136 |
| 10 | 11 | 86–134 | 81–139 | 77–143 | 73–147 | 65–155 |
| 11 | 11 | 100–153 | 96–157 | 91–162 | 87–166 | 78–175 |
| 2 | 12 | 5–25 | 4–26 | – | – | – |
| 3 | 12 | 11–37 | 10–38 | 8–40 | 7–41 | – |
| 4 | 12 | 19–49 | 17–51 | 15–53 | 13–55 | – |
| 5 | 12 | 28–62 | 26–64 | 23–67 | 21–69 | 16–74 |
| 6 | 12 | 38–76 | 35–79 | 32–82 | 30–84 | 24–90 |
| 7 | 12 | 49–91 | 46–94 | 42–98 | 10–100 | 33–107 |
| 8 | 12 | 52–106 | 58–110 | 53–115 | 51–117 | 43–125 |
| 9 | 12 | 75–123 | 71–127 | 66–132 | 63–135 | 55–143 |
| 10 | 12 | 89–141 | 84–146 | 79–151 | 76–154 | 67–163 |
| 11 | 12 | 104–160 | 99–165 | 79–170 | 90–174 | 81–183 |
| 12 | 12 | 120–180 | 115–185 | 109–191 | 105–195 | 95–205 |
| 2 | 13 | 5–27 | 4–28 | 3–29 | – | – |
| 3 | 13 | 12–39 | 10–41 | 8–43 | 7–44 | – |
| 4 | 13 | 20–52 | 18–545 | 15–57 | 13–59 | 10–62 |
| 5 | 13 | 30–65 | 27–68 | 24–71 | 22–73 | 17–78 |
| 6 | 13 | 40–80 | 37–83 | 33–87 | 31–89 | 25–95 |
| 7 | 13 | 52–95 | 48–99 | 44–103 | 41–106 | 34–113 |
| 8 | 13 | 64–112 | 60–116 | 56–120 | 53–123 | 45–131 |
| 9 | 13 | 78–129 | 73–134 | 68–139 | 65–142 | 56–151 |

(continued)

**Table A15** (continued)

| | | Two-tailed probability (P) | | | | |
|---|---|---|---|---|---|---|
| $n_1$ | $n_2$ | 0.1 | 0.05 | 0.02 | 0.01 | 0.001 |
| 10 | 13 | 92–148 | 88–152 | 82–158 | 79–161 | 69–171 |
| 11 | 13 | 108–167 | 103–172 | 97–178 | 93–182 | 83–192 |
| 12 | 13 | 125–187 | 119–193 | 113–199 | 109–203 | 98–214 |
| 13 | 13 | 142–209 | 136–215 | 130–221 | 125–226 | 114–237 |
| 2 | 14 | 6–28 | 4–30 | 3–31 | – | – |
| 3 | 14 | 13–41 | 11–43 | 8–46 | 7–47 | – |
| 4 | 14 | 21–55 | 19–57 | 16–60 | 14–62 | 10–66 |
| 5 | 14 | 31–69 | 28–72 | 25–75 | 22–78 | 17–83 |
| 6 | 14 | 42–84 | 38–88 | 34–92 | 32–94 | 26–100 |
| 7 | 14 | 54–100 | 50–104 | 45–109 | 43–111 | 35–119 |
| 8 | 14 | 67–117 | 62–122 | 58–126 | 54–130 | 46–138 |
| 9 | 14 | 81–135 | 76–140 | 71–145 | 67–149 | 58–158 |
| 10 | 14 | 96–154 | 91–159 | 85–165 | 81–169 | 71–179 |
| 11 | 14 | 112–174 | 106–180 | 100–186 | 96–190 | 85–201 |
| 12 | 14 | 129–195 | 123–201 | 116–208 | 112–212 | 100–224 |
| 13 | 14 | 147–217 | 141–223 | 134–2301 | 129–235 | 116–248 |
| 14 | 14 | 166–240 | 160–246 | 152–254 | 147–259 | 134–272 |
| 2 | 15 | 6–30 | 4–32 | 3–33 | – | – |
| 3 | 15 | 13–44 | 11–46 | 9–48 | 8–49 | – |
| 4 | 15 | 22–58 | 20–60 | 17–63 | 15–65 | 10–70 |
| 5 | 15 | 33–72 | 29–76 | 26–79 | 23–82 | 18–87 |
| 6 | 15 | 44–88 | 40–92 | 36–96 | 33–99 | 26–106 |
| 7 | 15 | 56–105 | 52–109 | 47–114 | 44–117 | 36–125 |
| 8 | 15 | 69–123 | 65–127 | 60–132 | 56–136 | 47–145 |
| 9 | 15 | 84–131 | 79–146 | 73–152 | 69–156 | 60–165 |
| 10 | 15 | 99–161 | 94–166 | 88–172 | 84–176 | 73–187 |
| 11 | 15 | 116–181 | 110–187 | 103–194 | 99–198 | 87–210 |
| 12 | 15 | 133–203 | 127–209 | 120–216 | 115–221 | 103–233 |
| 13 | 15 | 152–225 | 145–232 | 138–239 | 133–221 | 119–258 |
| 14 | 15 | 171–249 | 164–256 | 156–264 | 151–269 | 137–283 |
| 15 | 15 | 192–273 | 184–281 | 176–289 | 171–294 | 156–309 |

**Table A16** The Welch *t*-test (one-tailed and two-tailed)

| | *p*-level | | | | | | |
|---|---|---|---|---|---|---|---|
| One-tailed | 0.1 | 0.05 | 0.025 | 0.01 | 0.005 | 0.001 | 0.0005 |
| Two-tailed | 0.2 | 0.1 | 0.05 | 0.02 | 0.01 | 0.002 | 0.001 |
| Degrees of freedom | | | | | | | |
| 1 | 3.078 | 6.314 | 12.706 | 31.821 | 63.656 | 318.289 | 636.578 |
| 2 | 1.886 | 2.92 | 4.303 | 6.965 | 9.925 | 22.328 | 31.6 |
| 3 | 1.638 | 2.353 | 3.182 | 4.541 | 5.841 | 10.214 | 12.924 |
| 4 | 1.533 | 2.132 | 2.776 | 3.747 | 4.604 | 7.173 | 8.61 |
| 5 | 1.476 | 2.015 | 2.571 | 3.365 | 4.032 | 5.894 | 6.869 |
| 6 | 1.44 | 1.943 | 2.447 | 3.143 | 3.707 | 5.208 | 5.959 |
| 7 | 1.415 | 1.895 | 2.365 | 2.998 | 3.499 | 4.785 | 5.408 |
| 8 | 1.397 | 1.86 | 2.306 | 2.896 | 3.355 | 4.501 | 5.041 |
| 9 | 1.383 | 1.833 | 2.262 | 2.821 | 3.25 | 4.297 | 4.781 |
| 10 | 1.372 | 1.812 | 2.228 | 2.764 | 3.169 | 4.144 | 4.587 |
| 11 | 1.363 | 1.796 | 2.201 | 2.718 | 3.106 | 4.025 | 4.437 |
| 12 | 1.356 | 1.782 | 2.179 | 2.681 | 3.055 | 3.93 | 4.318 |
| 13 | 1.35 | 1.771 | 2.16 | 2.65 | 3.012 | 3.852 | 4.221 |
| 14 | 1.345 | 1.761 | 2.145 | 2.624 | 2.977 | 3.787 | 4.14 |
| 15 | 1.341 | 1.753 | 2.131 | 2.602 | 2.947 | 3.733 | 4.073 |
| 16 | 1.337 | 1.746 | 2.12 | 2.583 | 2.921 | 3.686 | 4.015 |
| 17 | 1.333 | 1.74 | 2.11 | 2.567 | 2.898 | 3.646 | 3.965 |
| 18 | 1.33 | 1.734 | 2.101 | 2.552 | 2.878 | 3.61 | 3.922 |
| 19 | 1.328 | 1.729 | 2.093 | 2.539 | 2.861 | 3.579 | 3.883 |
| 20 | 1.325 | 1.725 | 2.086 | 2.528 | 2.845 | 3.552 | 3.85 |
| 21 | 1.323 | 1.721 | 2.08 | 2.518 | 2.831 | 3.527 | 3.819 |
| 22 | 1.321 | 1.717 | 2.074 | 2.508 | 2.819 | 3.505 | 3.792 |
| 23 | 1.319 | 1.714 | 2.069 | 2.5 | 2.807 | 3.485 | 3.768 |
| 24 | 1.318 | 1.711 | 2.064 | 2.492 | 2.797 | 3.467 | 3.745 |
| 25 | 1.316 | 1.708 | 2.06 | 2.485 | 2.787 | 3.45 | 3.725 |
| 26 | 1.315 | 1.706 | 2.056 | 2.479 | 2.779 | 3.435 | 3.707 |
| 27 | 1.314 | 1.703 | 2.052 | 2.473 | 2.771 | 3.421 | 3.689 |
| 28 | 1.313 | 1.701 | 2.048 | 2.467 | 2.763 | 3.408 | 3.674 |
| 29 | 1.311 | 1.699 | 2.045 | 2.462 | 2.756 | 3.396 | 3.66 |
| 30 | 1.31 | 1.697 | 2.042 | 2.457 | 2.75 | 3.385 | 3.646 |
| 60 | 1.296 | 1.671 | 2 | 2.39 | 2.66 | 3.232 | 3.46 |
| 120 | 1.289 | 1.658 | 1.98 | 2.358 | 2.617 | 3.16 | 3.373 |
| 1000 | 1.282 | 1.646 | 1.962 | 2.33 | 2.581 | 3.098 | 3.3 |
| ∞ | 1.282 | 1.645 | 1.96 | 2.326 | 2.576 | 3.091 | 3.291 |

# Index




S. Belciug, *The Hospital Manager's Guide to Artificial Intelligence*, Intelligent Systems Reference Library 263, https://doi.org/10.1007/978-3-031-80314-7

**E**



**F**



**G**



**I**



**K**



**L**



**M**



**N**



**O**

**V**



**W**

The manufacturer's authorised representative in the EU is Springer Nature Customer Service Centre GmbH, Europaplatz 3, 69115 Heidelberg, Germany. If you have any concerns regarding our products, please contact ProductSafety@springernature.com

Printed and bound by CPI Group (UK) Ltd, Croydon, CR0 4YY
30/09/2026
02225586-0005